Compiled by Chinese Academy of Tropical Agricultural Sciences (CATAS) and Chinese Society for Tropical Crops (CSTC)
A Series of Books for Field Guide to Common Plants in FSM

General Editor: Liu Guodao

Field Guide to Medicinal Plants in the Federated States of Micronesia

Editors in Chief: Wang Qinglong Gu Wenliang

China Agricultural Science and Technology Press

图书在版编目（CIP）数据

密克罗尼西亚联邦药用植物图鉴 = Field Guide to Medicinal Plants in the Federated States of Micronesia / 王清隆，顾文亮主编 . —北京：中国农业科学技术出版社，2021.5

（密克罗尼西亚常见植物图鉴系列丛书 / 刘国道主编）

ISBN 978-7-5116-5289-8

Ⅰ . ①密… Ⅱ . ①王… ②顾… Ⅲ . ①药用植物—密克罗尼西亚联邦—图集 Ⅳ . ① S759.82-64

中国版本图书馆 CIP 数据核字（2021）第 068616 号

责任编辑 徐定娜
责任校对 贾海霞
责任印制 姜义伟 王思文

出 版 者 中国农业科学技术出版社
北京市中关村南大街 12 号 邮编：100081
电 话 （010）82105169（编辑室）（010）82109702（发行部）
（010）82109709（读者服务部）
传 真 （010）82109707
网 址 http://www.castp.cn
发 行 各地新华书店
印 刷 者 北京科信印刷有限公司
开 本 787 mm × 1 092 mm 1 /16
印 张 9.5
字 数 416 千字
版 次 2021 年 5 月第 1 版 2021 年 5 月第 1 次印刷
定 价 108.00 元

About the Author

Dr. Liu Guodao, born in June 1963 in Tengchong City, Yunnan province, is the incumbent Vice President of Chinese Academy of Tropical Agricultural Sciences (CATAS). Being a professor and PhD tutor, he also serves as the Director-General of the China-Republic of the Congo Agricultural Technology Demonstration Center.

In 2007, he was granted with his PhD degree from the South China University of Tropical Agriculture, majoring in Crop Cultivation and Farming.

Apart from focusing on the work of CATAS, he also acts as a tutor of PhD candidates at Hainan University, Member of the Steering Committee of the FAO Tropical Agriculture Platform (TAP), Council Member of the International Rubber Research and Development Board (IRRDB), Chairman of the Chinese Society for Tropical Crops, Chairman of the Botanical Society of Hainan, Executive Director of the Chinese Grassland Society and Deputy Director of the National Committee for the Examination and Approval of Forage Varieties and the National Committee for the Examination and Approval of Tropical Crop Varieties.

He has long been engaged in the research of tropical forage. He has presided over 30 national, provincial and ministerial-level projects: namely the "National Project on Key Basic Research (973 Program)" and international cooperation projects of the Ministry of Science and Technology, projects of the National Natural Science Foundation of China, projects of the International Center for Tropical Agriculture in Colombia and a bunch of projects sponsored by the Ministry of Agriculture and Rural Affairs (MARA) including the Talent Support Project, the "948" Program and the Infrastructure Project and Special

Scientific Research Projects of Public Welfare Industry.

He has published more than 300 monographs in domestic and international journals such as "New Phytologist" "Journal of Experimental Botany" "The Rangeland Journal" "Acta Prataculturae Sinica" "Acta Agrestia Sinica" "Chinese Journal of Tropical Crops", among which there nearly 20 were being included in the SCI database. Besides, he has compiled over 10 monographs, encompassing "Poaceae Plants in Hainan" "Cyperaceae Plants in Hainan" "Forage Plants in Hainan" "Germplasm Resources of Tropical Crops" "Germplasm Resources of Tropical Forage Plants" "Seeds of Tropical Forage Plants" "Chinese Tropical Forage Plant Resources". As the chief editor, he came out a textbook-*Tropical Forage Cultivation*, and two series of books-*Practical Techniques for Animal Husbandry in South China Agricultural Regions* (19 volumes) and *Practical Techniques for Chinese Tropical Agriculture "Going Global"* (16 volumes).

He has won more than 20 provincial-level and ministerial-level science and technology awards. They are the Team Award, the Popular Science Award and the First Prize of the MARA China Agricultural Science and Technology Award, the Special Prize of Hainan Natural Science Award, the First Prize of the Hainan Science and Technology Progress Award and the First Prize of Hainan Science and Technology Achievement Transformation Award.

He developed 23 new forage varieties including Reyan No. 4 King grass. He was granted with 6 patents of invention and 10 utility models by national patent authorities. He is an Outstanding Contributor in Hainan province and a Special Government Allowance Expert of the State Council.

Below are the awards he has won over the years: in 2020, "the Ho Leung Ho Lee Foundation Award for Science and Technology Innovation"; in 2018, "the High-Level Talent of Hainan province" "the Outstanding Talent of Hainan province" "the Hainan Science and Technology Figure"; in 2015, Team Award of "the China Agricultural Science and Technology Award" by the Ministry of Agriculture; in 2012, "the National Outstanding Agricultural Talents Prize" awarded by the Ministry of Agriculture and as team leader of the team award: "the Ministry of Agriculture Innovation Team" (focusing on the research of Tropical forage germplasm innovation and utilization); in 2010, the first-level candidate of the "515 Talent Project" in Hainan province; in 2005, "the Outstanding Talent of Hainan province"; in 2004, the first group of national-level candidates for the "New Century Talents Project" "the 4th Hainan Youth Science and Technology Award" "the 4th Hainan Youth May 4th Medal" "the 8th China Youth Science and Technology Award" "the Hainan Provincial International Science and Technology Cooperation Contribution Award"; in 2003, "a Cross-Century Outstanding Talent" awarded by the Ministry of Education; In 2001, "the 7th China Youth Science and Technology Award" of Chinese Association of Agricultural Science Societies, "the National Advanced Worker of Agricultural Science and Technology"; in 1993, "the Award for Talents with Outstanding Contributions after Returning to China" by the State Administration of Foreign Experts Affairs.

Dr. Wang Qinglong, born in September 1983 in Chengmai County, Hainan province, is a Professor Assistant in the Tropical Crops Genetic Resources Institute, Chinese Academy of Tropical Agricultural Sciences, mainly engaged in the research of plant taxonomy and resources. He has collected over 1,800 species of South China Medical Plants, released 4 new cultivars and over 50 newly combined and recorded cultivars in the world. He has accomplished several major projects, such as the projects of the National Natural Science Foundation of China, and published 31 research papers in "Phytotaxa" "Phytokeys" "Nordic Journal of Botany" "Journal of Tropical and Subtropical Botany", etc. He also serves as the Chief Editor or Deputy Chief Editor of many monographs, such as *Wild Plants in the South China Sea, Medicinal Cyperaceae Plants in Hainan,* etc. He won the title of "Excellent Lecturer" for Farmer Education and Training in Hainan province, and the First Prize of Hainan Science and Technology Progress Award in 2020.

A Series of Books for Field Guide to Common Plants in FSM

General Editor: Liu Guodao

Field Guide to Medicinal Plants in the Federated States of Micronesia

Editorial Board

The President

Palikir, Pohnpei

Federated States of Micronesia

FOREWORD

It is with great pleasure that I present this publication, "Agriculture Guideline Booklet" to the people of the Federated States of Micronesia (FSM).

The Agriculture Guideline Booklet is intended to strengthen the FSM Agriculture Sector by providing farmers and families the latest information that can be used by all in our communities to practice sound agricultural practices and to support and strengthen our local, state and national policies in food security. I am confident that the comprehensive notes, tools and data provided in the guideline booklets will be of great value to our economic development sector.

Special Appreciation is extended to the Government of the People's Republic of China, mostly the Chinese Academy of Tropical Agricultural Sciences (CATAS) for assisting the Government of the FSM especially our sisters' island states in publishing books for agricultural production. Your generous assistance in providing the practical farming techniques in agriculture will make the people of the FSM more agriculturally productive.

I would also like to thank our key staff of the National Government, Department of Resources and Development, the states' agriculture and forestry divisions and all relevant partners and stakeholders for their kind assistance and support extended to the team of Scientists and experts from CATAS during their extensive visit and work done in the FSM in 2018.

We look forward to a mutually beneficial partnership.

Sincerely,

David W. Panuelo

President

Preface

Claiming waters of over 3,000 square kilometers, the vast area where Pacific island countries nestle is home to more than 10,000 islands. Its location at the intersection of the east-west and north-south main traffic artery of Pacific wins itself geo-strategic significance. There are rich natural resources such as agricultural and mineral resources, oil and gas here. The relationship between the Federated States of Micronesia (hereinafter referred to as FSM) and China ushered in a new era in 2014 when Xi Jinping, President of China, and the leader of FSM decided to establish a strategic partnership on the basis of mutual respect and common development. Mr Christian, President of FSM, took a successful visit to China in March 2017 during which a consensus had been reached between the leaders that the traditional relationship should be deepened and pragmatic cooperation (especially in agriculture) be strengthened. This visit pointed out the direction for the development of relationship between the two countries. In November 2018, President Xi visited Papua New Guinea and in a collective meeting met 8 leaders of Pacific Island countries (with whom China has established diplomatic relation). China elevated the relationship between the countries into a comprehensive and strategic one on the basis of mutual respect and common development, a sign foreseeing a brand new prospect of cooperation.

The government of China launched a project aimed at assisting FSM in setting up demonstration farms in 1998. Until now, China has completed 10 agricultural technology cooperation projects. To answer the request of the government of FSM, Chinese Academy of Tropical Agricultural Sciences (hereinafter referred to as CATAS), directly affiliated with the

Ministry of Agriculture and Rural Affairs of China, was elected by the government of China to carry out training courses on agricultural technology in FSM during 2017—2018. The fruitful outcome is an output of training 125 agricultural backbone technicians and a series of popular science books which are entitled "Field Guide to Forages in the Federated States of Micronesia" "Field Guide to Flowers and Ornamental Plants in the Federated States of Micronesia" "Field Guide to Medicinal Plants in the Federated States of Micronesia" "Field Guide to Fruits and Vegetables in the Federated States of Micronesia" "Coconut Germplasm Resources in the Federated States of Micronesia" and "Field Guide to Plant Diseases, Insect Pests and Weeds in the Federated States of Micronesia".

In these books, 492 accessions of germplasm resources such as coconut, fruits, vegetables, flowers, forages, medical plants, and pests and weeds are systematically elaborated with profuse inclusion of pictures. They are rare and precious references to the agricultural resources in FSM, as well as a heart-winning project among China's aids to FSM.

Upon the notable moment of China-Pacific Island Countries Agriculture Ministers Meeting, I would like to send my sincere respect and congratulation to the experts of CATAS and friends from FSM who have contributed remarkably to the production of these books. I am firmly convinced that the exchange between the two countries on agriculture, culture and education will be much closer under the background of the publication of these books and Nadi Declaration of China-Pacific Island Countries Agriculture Ministers Meeting, and that more fruitful results will come about. I also believe that the team of experts in tropical agriculture mainly from the CATAS will make a greater contribution to closer connection in agricultural development strategies and plans between China and FSM, and closer cooperation in exchanges and capacity-building of agriculture staffs, in agricultural science and technology for the development of agriculture of both countries, in agricultural investment and trade, in facilitating FSM to expand industry chain and value chain of agriculture, etc.

Qu Dongyu

Director General

Food and Agriculture Organization of the United Nations

July 23, 2019

Located in the northern and central Pacific region, the Federated States of Micronesia (FSM) is an important hub connecting Asia and America. Micronesia has a large sea area, rich marine resources, good ecological environment, and unique traditional culture.

In the past 30 years since the establishment of diplomatic relations between China and FSM, cooperation in diverse fields at various levels has been further developed. Since the 18th National Congress of the Communist Party of China, under the guidance of Xi Jinping's thoughts on diplomacy, China has adhered to the fine diplomatic tradition of treating all countries as equals, adhered to the principle of upholding justice while pursuing shared interests and the principle of sincerity, real results, affinity, and good faith, and made historic achievements in the development of P.R. China-FSM relations.

The Chinese government attaches great importance to P.R. China-FSM relations and always sees FSM as a good friend and a good partner in the Pacific island region. In 2014, President Xi Jinping and the leader of the FSM made the decision to build a strategic partnership featuring mutual respect and common development, opening a new chapter of P.R. China-FSM relations. In 2017, FSM President Peter Christian made a successful visit to China. President Xi Jinping and President Christian reached broad consensuses on deepening the traditional friendship between the two countries and expanding practical cooperation between the two sides, and thus further promoted P.R. China-FSM relations. In 2018, Chinese President Xi Jinping and Micronesian President Peter Christian had a successful meeting again in PNG and made significant achievements, deciding to upgrade P.R. China-FSM relations to a new stage of Comprehensive Strategic Partnership, thus charting the course for future long-term development of P.R. China-FSM relations.

In 1998, the Chinese government implemented the P.R. China-FSM demonstration farm project in FSM. Ten agricultural technology cooperation projects have been completed, which has become the "golden signboard" for China's aid to FSM. From 2017 to 2018, the Chinese Academy of Tropical Agricultural Sciences (CATAS), directly affiliated with the Ministry of Agriculture and Rural Affairs, conducted a month-long technical training on pest control of coconut trees in FSM at the request of the Government of FSM. 125 agricultural managers, technical personnel and growers were trained in Yap, Chuuk, Kosrae and Pohnpei, and the biological control technology demonstration of the major dangerous pest, Coconut Leaf Beetle, was carried out. At the same time, the experts took advantage of the spare time of the training course and spared no effort to carry out the preliminary evaluation of the investigation and utilization of agricultural resources, such as coconut, areca nut, fruit tree, flower, forage, medicinal plant, melon and vegetable, crop disease, insect pest and weed diseases, in the field in conjunction with Department of Resources and Development of FSM and the vast number of trainees, organized and compiled a series of popular science books, such as "Field Guide to Forages in the Federated States of Micronesia" "Field Guide to Flowers and Ornamental Plants in the Federated States of Micronesia" "Field Guide to Medicinal Plants in the Federated States of Micronesia" "Field Guide to Fruits and Vegetables in the Federated States of Micronesia" "Coconut Germplasm Resources in the Federated States of Micronesia" and "Field Guide to Plant Diseases, Insect Pests and Weeds in the Federated States of Micronesia".

The book introduces 37 kinds of coconut germplasm resources, 60 kinds of fruits and vegetables, 91 kinds of angiosperm flowers as well as 13 kinds of ornamental pteridophytes, 100 kinds of forage plants, 117 kinds of medicinal plants, 74 kinds of crop diseases, pests and weed diseases, in an easy-to-understand manner. It is a rare agricultural resource illustration in FSM. This series of books is not only suitable for the scientific and educational workers of FSM, but also it is a valuable reference book for industry managers, students, growers and all other people who are interested in the agricultural resources of FSM.

This series is of great significance for it is published on the occasion of the 30th anniversary of the establishment of diplomatic relations between the People's Republic of

China and FSM. Here, I would like to pay tribute to the experts from CATAS and the friends in FSM who have made outstanding contributions to this series of books. I congratulate and thank all the participants in this series for their hard and excellent work. I firmly believe that based on this series of books, the agricultural and cultural exchanges between China and FSM will get closer with each passing day, and better results will be achieved more quickly. At the same time, I firmly believe that the Chinese Tropical Agricultural Research Team, with CATAS as its main force, will bring new vigour and make new contributions to promoting the in-depth development of the strategic partnership between the People's Republic of China and the Federated States of Micronesia, strengthening solidarity and cooperation between P.R. China and the developing countries, and the P.R. China-FSM joint pursuit of the Belt and Road initiative and building a community with a shared future for the humanity.

Ambassador Extraordinary & Plenipotentiary of
the People's Republic of China to
the Federated States of Micronesia
May 23, 2019

Foreword

Chinese herbal medicine, a unique kind of medicines adopted by doctors of traditional Chinese medicine for the prevention and treatment of diseases and for health care, is a valuable asset acquired through fighting against various diseases for thousands of years and through personal experiences of hundreds of millions of people during the years. Chinese herbal medicine has gained increasing attention and recognition from the international community for its advantages such as natural origins, no pollution, good curative effects and low cost.

The Federated States of Micronesia (FSM), located in the central Pacific, is an important hub connecting Asia and America. The FSM area boasts a large sea area and abundant marine resources. The special geographical locations of the islands under the influence of tropical monsoon, enjoy a warm climate, plentiful sunshine and rainfall and rich plant resources.

In 2018, in order to implement the foreign aid training program supported by the Ministry of Commerce, PRC, a scientist team from Chinese Academy of Tropical Agricultural Sciences (CATAS) successfully conducted trainings on the prevention and control of coconut tree diseases and insect pests in the FSM, with participants of 125 agricultural officials, technical personnel and farm holders in the four states of Yap, Chuuk, Kosrae and Pohnpei, and demonstrated biological control techniques for *Brontispa longissima*, a detrimental vermin of coconut trees. During this period, CATAS scientists, by taking advantage of their spare time, made a full-coverage filed survey and preliminary evaluation on the agriculturally biological resources including coconut palms, fruit plants, flower plants, forage plants, medicinal plants, cucurbits and vegetables and crop diseases, insect pests and weeds, through cooperating with the FSM Ministry of Agriculture and Resources and trainees. The field survey team was rewarded with great success in collecting various biological resource data in spite of varied difficulties. On the basis of field survey they compiled a series of interesting works with popular science in nature, including "Field Guide to Forages in the Federated States of Micronesia" "Field Guide to Flowers and Ornamental Plants in the Federated States of Micronesia" "Field Guide to Medicinal Plants in the Federated States of Micronesia" "Field Guide to Fruits and Vegetables in the Federated States of Micronesia" "Coconut Germplasm Resources in the Federated States of Micronesia" and "Field Guide to Plant Diseases, Insect Pests and Weeds in the Federated States of Micronesia".

Compilation of this monograph in Chinese was not done until consultation and textual research had been made on extensively collected literature against field survey records. Some 117 kinds of medicinal plants in the FSM are documented into this monograph, including fern, dicotyledon and monocotyledon. Each of the written contents includes Chinese name (Chinese version), scientific name, common name in English (English version), property, habitat distribution, medicinal part, efficacies, indications, usage and dosage, accompanied by beautiful illustrative pictures. It is a monograph comprehensively introducing the resources and use value of medicinal plants in the FSM, a valuable reference book for scientists, staff and students at universities or colleges, hospital doctors, related government officers, farmers, and all other people interested in medicinal plant resources in the FSM. It is also part of the achievements of an agricultural project for the public interests of the local people in the form of Chinese government's aid to the FSM government.

For better understanding and convenience for use, the original Chinese manuscript is translated into English version. Both the Chinese and English version shares the same contents with one exception, i.e., no Chinese name to each entry in the English version but a common name in English instead.

刘国道

General Editor

Vice President of Chinese Academy of Tropical Agricultural Sciences

March 22, 2019

Contents

● Psilotaceae

Psilotum nudum (L.) Beauv.

Common name: Whist fern

Property: Epiphytic herb.

Habitat Distribution: Grows on rocks or attaches to tree trunks; widely distributed in tropical and subtropical regions.

Medicinal Part: Whole herb.

Efficacies: To relieve rheumatic conditions, to promote the flow of blood and to arrest bleeding.

Indications: Rheumatic arthalgia, rubella, amenorrhea, hematemesis, traumatic injury.

Usage and Dosage: Oral administration: make decoction, 25 to 50 g; or grind into fine powder or infuse in wine.

● Lycopodiaceae

Palhinhaea cernua (L.) Vasc. et Franco

Common name: Nodding clubmoss

Property: Herb.

Habitat Distribution: Grows in wildernesses, wastelands and on roadsides;distributed in tropical and subtropical Asia, Oceania, Central and South America.

Medicinal Part: Whole herb.

Efficacies: To eliminate rheumatism, remove obstruction in meridians and collaterals.

Indications: Rheumatic contrcture, hepatitis, dysentery, rubella, acute conjunctivitis, hematemesis, non-traumatic hemorrhage, hematochezia, traumatic injury, burns.

Usage and Dosage: Oral administration: make decoction, 6 to 15 g (fresh herb 50 to 100 g). External use: wash with decoction or grind into fine powder and apply.

● Ophioglossaceae

Ophioglossum vulgatum L.

Common name: Adder's tongue, common Adder's tongue

Property: Herb.

Habitat Distribution: Grows under forests; widely distributed in Europe, Asia and America.

Medicinal Part: Whole herb.

Efficacies: To remove heat to cool blood, to remove toxic substances to relieve pain.

Indications: Cough due to pathogenic lung heat, pulmonary abscess, hematemesis due to pulmonary tuberculosis, infantile convulsion due to high fever, painful conjunctival congestion, stomachache, furuncles, snake and insect bites, and traumatic injury.

Usage and Dosage: Oral administration: make decoction, 9 to 15g, or together with external use: smash with appropriate amount and apply on the affected part.

● Lygodiaceae

Lygodium microphyllum (Cav.) R. Br

Common name: Small-leaf climbing fern

Property: Rampant climbing herb.

Habitat Distribution: Grows on roadsides and in streamside shrubs with sufficient sunlight; distributed in China, southern India, Myanmar, Philippines and South Pacific island countries.

Medicinal Part: Whole herb and spores.

Efficacies: To remove damp-heat and relieve dysuria, to relax muscles and tendons and remove obstruction from meridians, to relieve stranguria, and to arrest bleeding.

Indications: Edema, hepatitis, stranguria, dysentery, hemafacia, rheumatic numbness, traumatic bleeding.

Usage and Dosage: Oral administration: Decocted in cheesecloth, 5 to 9 g; or grind into fine powder, 2 to 3 g each time.

● Pteridaceae

Pteris vittata L.

Common name: Centipedal brake, Laker brake

Property: Herb.

Habitat Distribution: Grows on walls or gaps; distributed in the Old World and other tropical and subtropical regions of the world.

Medicinal Part: Whole herb or rhizome.

Efficacies: To relieve rheumatic conditions or expel wind and to promote the flow of blood, and to counteract toxicity and kill parasites.

Indications: Influenza, dysentery,rheumatalgia, traumatic injury, centipede bites, scabies.

Usage and Dosage: Oral administration: make decoction, 6 to 12g. External use: wash with decoction or smash and apply.

● Parkeriaceae

Ceratopteris thalictroides (L.) Brongn.

Common name: Floating fern, Oriental water ferm

Property: Herb.

Habitat Distribution: Grows in muds of ponds, paddy fields or gutters; widely distributed in the tropical and subtropical regions of the world as well as Japan.

Medicinal Part: Whole herb.

Efficacies: To dissipate blood stasis and to remove toxins, to relieve cough, to resolve phlegm, to relieve dysentery, and to arrest bleeding.

Indications: Fetal toxicity(infantile carbuncle), phlegm accumulation, traumatic injury,cough, dysentery, turbid stranguria; traumatic bleeding (external use).

Usage and Dosage: Oral administration: make decoction, 15 to 30 g. External use: smash with appropriate amount and apply.

● Blechnaceae

Blechnum orientale L.

Common name: Oriental Blechnum

Property: Herb.

Habitat Distribution: Grows beside dank gutters, on pit edges, in hillside shrubs or under sparse forests; distributed in China, India, Sri Lanka, Southeast Asia, Japan and Polynesia.

Medicinal Part: Rhizome.

Efficacies: To remove toxic heat, to promote the flow of blood, to arrest bleeding, and to expel intestinal parasitic worms.

Indications: Cold, headache, parotitis, carbuncle, traumatic injury, epistaxis, hematemesis, metrorrhagia, leucorrhoea, and entorozoon.

Usage and Dosage: Oral administration: make decoction, 6 to 15 g (up to 60 g in a large dosage). External use: smash with appropriate amount and apply; or grind into fine powder and apply.

● Polypodiaceae

Phymatosorus scolopendria (Burm.) Pic. Serm

Common name: Common phymotodes

Property: Epiphytic herb.

Habitat Distribution: Grows on raw stones or attaches to tree trunks; distributed in Japan,Indo-China Peninsula, Philippines, Malaysia, Thailand, India, Sri Lanka, New Guinea, tropical Australia, tropical Africa and Polynesia.

Medicinal Part: Rhizome.

Efficacies: To Promote blood circulation to cause subsistence of swelling,to promote the reunion of fractured bones.

Indications: Traumatic injury, traumatic bleeding, scalds, ureteralgia.

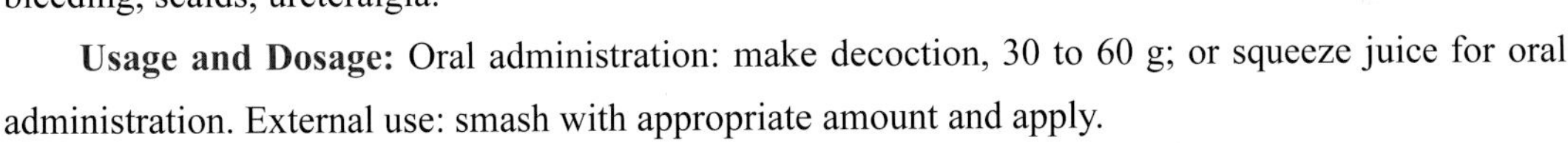

Usage and Dosage: Oral administration: make decoction, 30 to 60 g; or squeeze juice for oral administration. External use: smash with appropriate amount and apply.

● Lauraceae

Cassytha filiformis L.

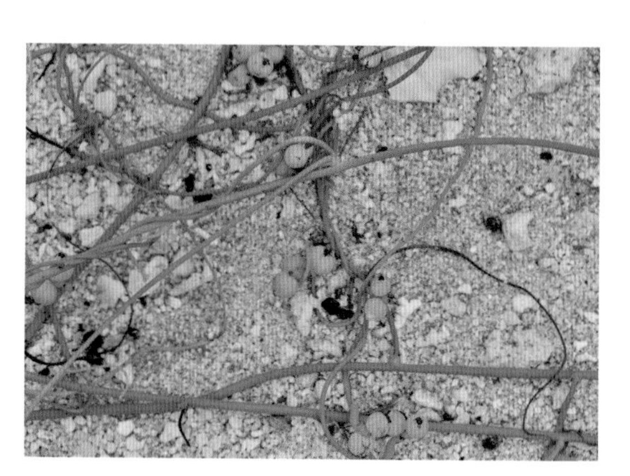

Common name: Filiform Cassytha

Property: Liana.

Habitat Distribution: Grows on seaside, hillsides and wild shrubs; distributed in tropical Asia, Africa and Australia.

Medicinal Part: Whole herb.

Efficacies: To dissipate dampness to subdue swelling, to relieve stranguria to promote diuresis.

Indications: Cold fever, malaria, acute jaundice hepatitis, hemoptysis, epistaxis, hematuria, lithangiuria, nephritic edema;eczema and multiple furuncles (external use).

Usage and Dosage: Oral administration: make decoction, 9 to 15 g. External use: smash with appropriate amount and apply, or wash with decoction.

Cinnamomum verum Presl

Common name: Ceylon cinnamon

Property: Evergreen arbor.

Habitat Distribution: Native to Sri Lanka; cultivated in Micronesia.

Medicinal Part: Bark.

Efficacies: To warm the spleen and stomach to tonify the kidney, and to dispel cold to alleviate pain.

Indications: Crymodynia of knees and loin, deficiency-cold stomachache, chronic indigestion,abdominalgia and vomiting and diarrhea, chill amenorrhea.Twigs of Ceylon cinnamon can promote sweating, help to expellpathogenic factors from muscles and skin, and warmthe meridians for the treatment of affection of exogenous wind-cold and aching of shoulder and arm. The decoction of twigs has significantly antibacterial effects on *Staphylococcus aureus,Salmonella typhi and Hominis mycobacterium tuberculosis.* The seeds can be used for the treatment of deficiency-cold stomachache.

Usage and Dosage: Oral administration: make decoction, 1.5 to 4.5 g; or use in pills or powder. External use: grind into fine powder and apply or infuse in wine for inunction.

● Myristicaceae

Myristica fragrans Houtt.

Common name: Nutmeg

Property: Small tree.

Habitat Distribution: Native to the Maluku Islands; widely cultivated in tropical regions.

Medicinal Part: Seed.

Efficacies: To warm the spleen and stomach, to check upward rise of *Qi* (gas), to promote digestion, and to check diarrhea.

Indications: Distending painin chest and abdomen, diarrhea of deficiency type and cold dysentery, vomiting, and dyspepsia; parasitic worms and rheumatism (external use).

Usage and Dosage: Oral administration: make decoction, 1.5to 6 g; or use in pills or powder.

● Nepenthaceae

Nepenthes mirabilis (Lour.) Merr.

Common name: Common nepenthes

Property: Standing upright or climbing herb.

Habitat Distribution: Grows in the shrubs, grasslands or under the forests of marshlands, roadsides, mountainside and mountaintops at an altitude from 50 to 400 meters; adaptable to various environments; widely distributed from the Indo-China Peninsula to the North Oceania.

Medicinal Part: Whole herb.

Efficacies: To clear away rheumatic pneumonia (heat in the lung), and quench thirsty, to keep the normal flow of body fluids, and detoxification.

Indications: Cough due to dryness in the lung, pertussis, jaundice, stomachache, dysentery, edema, carbuncle, insect bites.

Usage and Dosage: Oral administration: make decoction, 25 to 50 g (fresh herb 50 to 100 g). External use: smash and apply.

● Piperaceae

Piper betle L.

Common name: Betel pepper

Property: Climbing liana.

Habitat Distribution: Commonly cultivated; distributed in India, Sri Lanka, Vietnam, Malaysia, Indonesia, the Philippines and Madagascar.

Medicinal Part: Stem and leaf.

Efficacies: To remove rheumatic conditions, to promote the flow of *qi* to resolve phlegm, to suside swelling and to relieve itching.

Indications: Wind-cold Cough, bronchial asthma, rheumatic osteodynia, cold gastralgia, gestational edema; skin eczema and tinea pedis (external use).

Usage and Dosage: Oral administration: make decoction, 3 to 9 g. External use: make decoction with appropriate amount and wash at a suitable temperature; soak for tinea pedis.

Piper methysticum Forst

Common name: Kava Pepper

Property: Perennial upright shrub.

Habitat Distribution: Grows under forests; distributed in the South Pacific Islands.

Medicinal Part: Root and rhizome.

Efficacies: To regulate neurotransmitters bi-directionally, to relieve dysphoria and depression, sedative hypnosis to induce local anesthesia, and to arrest convulsion.

Indications: Chronic fatigue syndrome, fibromyalgia syndrome, insomnia, nephritis, cystitis, vaginitis and urethritis.

Usage and Dosage: Oral administration: make decoction, 1.5 to 3g; or use in pills or powder. External use: grind into fine powder and apply or put onto a plaster for paste.

Piper nigrum L.

Common name: Pepper

Property: Woody climbing liana.

Habitat Distribution: Native to Southeast Asia; widely grown in tropical regions; cultivated in Micronesia.

Medicinal Part: Fruit.

Efficacies: To eliminate phlegm and to remove toxicity.

Indications: Cold phlegm and dyspepsia, abdominal crymodynia (epigastric pain with cold sensation), regurgitation, vomiting fluid, diarrhea, cold dysentery, food poisoning.

Usage and Dosage: Oral administration: make decoction, 1.5 to 3g; or use in pills or powder. External use: grind into fine powder and apply or put onto a plaster for paste.

Piper sarmentosum Roxb.

Common name: Runner pepper

Property: Perennial creeping herb of rooting by section.

Habitat Distribution: Grows under the forests or on the wetlands near villages; distributed in China, India, Vietnam, Malaysia, the Philippines, Indonesia and Papua New Guinea.

Medicinal Part: Root or fruit.

Efficacies: To eliminate malaria, beriberi toothache, hemorrhoids.

Indications: Radical cureof rheumatic osteodysnia, traumatic injury cough due to wind-cold, gestational and postpartum edema;infructescence for treatment of toothache, stomachache, abdominal distension and anorexia.

Usage and Dosage: Oral administration: make decoction (fresh herb 10 to 15 g). External use: smash and apply or wash with decoction.

● Capparidaceae

Arivela viscosa (L.) Raf.

Common name: Yellow flower, spider flower, Asian spider flower

Property: Upright annual herb.

Habitat Distribution: Quite different ecological environment; commonly seen in wastelands, fields and on roadsides in dry areas, native to the Old Tropics, now a kind of medicinal plant and weed found in tropical and subtropical regions of the world.

Medicinal Part: Whole herb.

Efficacies: To remove blood stasis to subsidize swelling, to remove necrotic tissue while promote tissue regeneration.

Indications: Traumatic swelling and pain, over-strained lumbago, uletrous sore.

Usage and Dosage: External use:wash with decoction of whole herb, and grind into fine powderand apply on the affected part.

Crateva religiosa G. forst.

Common name: Sacred Barma

Property: Shrub or arbor.

Habitat Distribution: Grows on roadsides and in sparse forests; distributed in Japan, Australia, Southeast Asia and the South Pacific Islands.

Medicinal Part: Bark, leaf, sap and flower.

Efficacies: To subside fever, to arrest diarrhea, to eliminate dampness in the spleen and stomach.

Indications: Diarrhea pain, rheumatism, earache, astriction,gallbladder function normalization.

Usage and Dosage: Root and bark: 9 to 15 g; leaf: oral administration; make decoction, 4.5 to 10 g. External use: smash and apply or wash with decoction.

Abstinence: 1. Contraindicated in pregnancy; 2.Food abstinence: with chicken at the same time.

● Molluginaceae

Mollugo stricta L.

Common name: Carpet weed

Property: Annual diffusing herb.

Habitat Distribution: Grows in open wastelands, farmlands and coastal sands; distributed in tropical and subtropical regions of Asia.

Medicinal Part: Whole herb.

Efficacies: To remove internal toxic-heat.

Indications: Abdominal pain due to diarrhea, rubellaacute conjunctivitis and venomous snake bite.

Usage and Dosage: Oral administration: make decoction, 9 to 30 g. External use: smash with appropriate amount of fresh herb and stuff into nose or apply on the affected part.

Amaranthaceae

Alternanthera sessilis (L.) DC.

Common name: Sessile joyweed

Property: Perennial herb.

Habitat Distribution: Grows in the grass slopes, ditches, fields or swamps near villages and wet places on seaside; also distributed in China, India, Myanmar, Vietnam, Malaysia and Philippines.

Medicinal Part: Whole herb.

Efficacies: To dissipate blood stasis and remove toxicity, to purge internal fire and reduce fever.

Indications: Toothache, dysentery, and bloody defecation.

Usage and Dosage: Oral administration: make decoction,15 to 30 g or squeeze juice with 60 to 120 g of fresh whole herb, stew and take warm decoction. External use: smash with appropriate amount of fresh whole herb and apply or make thick decoction and wash the affected part.

Celosia argentea L.

Common name: Semen celosia

Property: Annual herb.

Habitat Distribution: Grows in plains, fields, on hills andhillsides; distributed in China, North Korea, Japan, Russia, India, Vietnam, Myanmar, Thailand, the Philippines, Malaysia and tropical Africa.

Medicinal Part: Seed or whole herb.

Efficacies: To clear liver-heat and improve vision, to descend blood pressure, and to clear heat to remove dampness in the liver and the gallbladder.

Indications: Eyes disorder due to pyretic toxicity of the liver,conjunctive congestion with nebula, optic atrophy, nebula, wind-heat headache manifested as headache with flushed face anddizziness, or splitting headache, conjunctival congestion and thirst, headache anddizziness, hypertension.

Usage and Dosage: Oral administration: make decoction or crush to extract juice with 30 to 60 g of fresh herb. External use: smash and apply.

Celosia cristata L.

Common name: Cockcomb

Property: Annual herb.

Habitat Distribution: Native to Africa, tropical America and India; widely cultivated all over the world.

Medicinal Part: Flower and seed.

Efficacies: To arrestbleeding, to remove heat from blood due to pathogenic invasion, and to relieve diarrhea.

Indications: Hematemesis, metrorrhagia, hemafacia, hemorrhoidal bleeding, multi-colored leukorrhagia, chronic dysentery.

Usage and Dosage: Oral administration: make decoction, 4.5 to 10 g; or use in pills or powder. External use: make decoction for fumigation and washing.

Cyathula prostrata (L.) Blume

Common name: Pasture weed

Property: Perennial herb.

Habitat Distribution: Grows in the shrubs of hillsides or on small riversides; distributed in China, Vietnam, India, Thailand, Myanmar, Malaysia, the Philippines, Africa and Oceania.

Medicinal Part: Whole herb.

Efficacies: To subdueswelling, to relieve pain, to expel gunshot bullet or shrapnel from lodged body parts,to remove various toxins.

Indications: Various venomous snake bites, hepatosplenomegaly, lodged bullet or shrapnel in body parts.

Usage and Dosage: Oral administration: make decoction, 30 to 60 g. External use: smash with appropriate amount and apply.

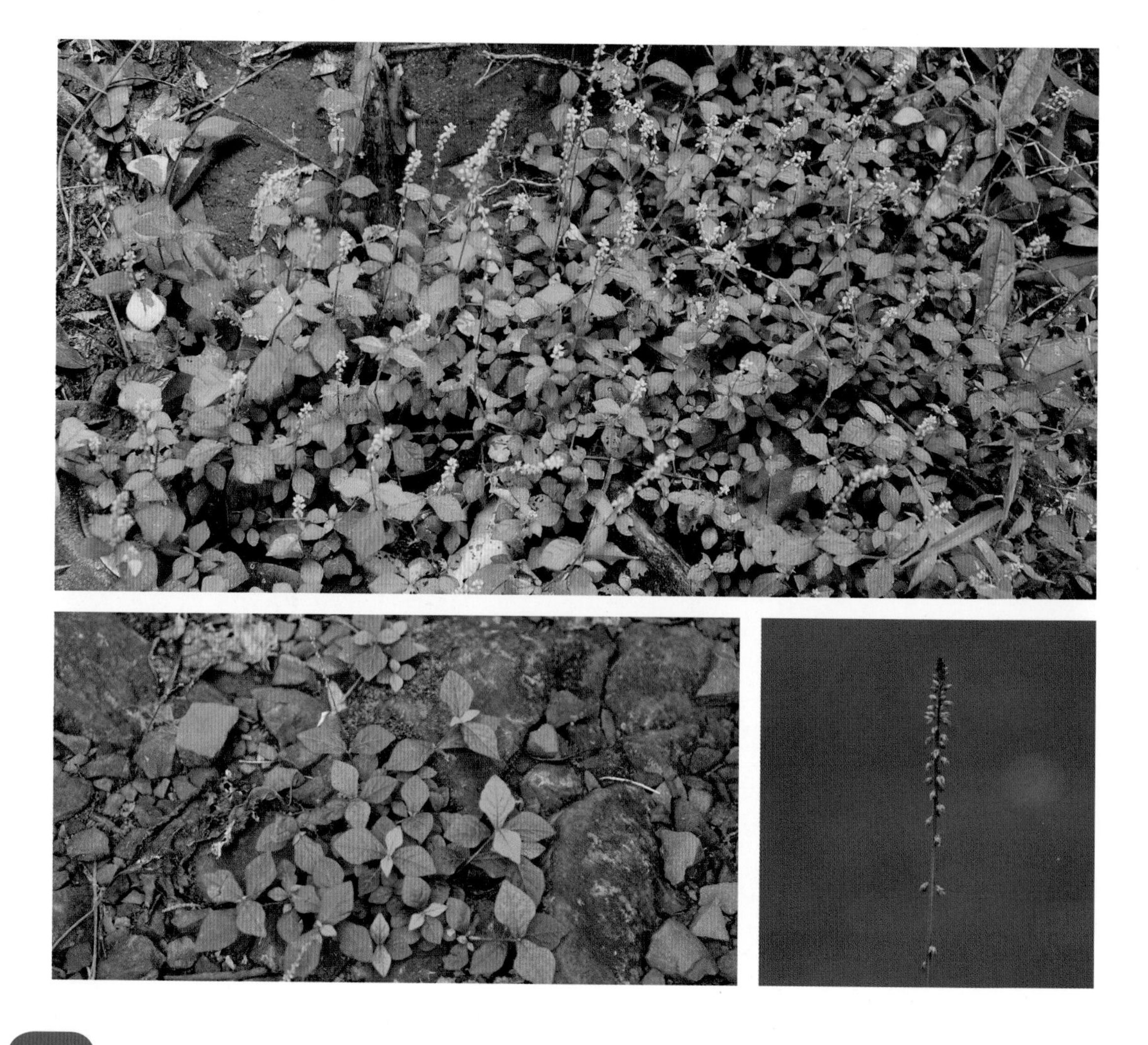

● Oxalidaceae

Averrhoa bilimbi L.

Common name: Bilimbi, cucumber tree

Property: Dungarunga.

Habitat Distribution: Grown in regions at an altitude from 120 to 1,200 meters; distributed in Malaysia, tropical Asia, India and China's Guangdong,Guangxi and Taiwan.

Medicinal Part: Leaf or flower.

Efficacies: To remove heat and promote saltivation,to promote diuresis to relieve stranguria.

Indications: Pyreticosis with polydipsia, anemopyrectic cough, sore throat, aphtha, dysuria, stony stranguria; leaf for pruritus, swelling, rheumatism, parotitis or rashes; flower for the treatment of thrush, cold and cough.

Usage and Dosage: Oral administration: make decoction, 30 to 60 g. External use: squeeze juice with appropriate amount and drop it into ear.

Oxalis corniculata L.

Common name: Woodsorrel

Property: Perennial herb.

Habitat Distribution: Grown in farmlands, wastelands or roadsides; distributed all over China.

Medicinal Part: Whole herb.

Efficacies: To clear internal heat to remove dampness, to cool blood for dispersing stasis, to subsidize swelling and removetoxical materials.

Indications: Diarrhea, dysentery, jaundice, gonorrhea, multi-coloured leukorrhea, measles, hematemesis, epistaxis, pharyngalgia, furuncle, carbuncle with swelling,scabies and tinea, dysentery, proctoptosis, traumatic injuries scald and burn.

Usage and Dosage: Oral administration:10 to 30 g; or grind into fine powder; or squeeze juice with fresh herb. External use: wash with decoction of appropriate amount; smash and apply; crush to extract juice for inunction or make decoction for mouthwash.

● Lythraceae

Pemphis acidula J. R. et Forst.

Common name: Pemphis

Property: Multi-branched shrub or dungarunga.

Habitat Distribution: Grown on seasides; distributed along the tropical coastsof the Eastern Hemisphere.

Medicinal Part: Twigs.

Efficacies: To remove phlegm and dampness to remove bloodstasis and relievepain.

Indications: Damp obstruction and blood stasis.

Usage and Dosage: Oral administration: make decoction, 5 to 10 g.

● Onagraceae

Ludwigia hyssopifolia (G. Don) Exell

Common name: Water primrose

Property: Upright annual herb.

Habitat Distribution: Grown in moist and sunny place along farm fields,ditches, river banks, pond-sides and wet grasslands; distributed in China, India, Sri Lanka, Myanmar, the Indo-China Peninsula through the Malay Peninsula to Philippines, Northern Australia, west to tropical Africa.

Medicinal Part: Whole herb.

Efficacies: To remove toxic-heat,to remove necrotic tissues andpromote tissue regeneration.

Indications: Cold, pharyngalgiascabies.

Usage and Dosage: Oral administration: make decoction, 10 to 30 g. External use: smash with appropriate amount and apply or make decoction for mouthwash.

● Haloragidaceae

Haloragis micrantha (Thunb.) R. Br.

Common name: Haloragia

Property: Perennial terrestrial herb.

Habitat Distribution: Grown among grasses of barren hills; distributed in China, Australia, New Zealand, Malaysia, India, Vietnam, Thailand, Japan and North Korea.

Medicinal Part: Whole herb.

Efficacies: To clear toxic-heat, to promote diuresisi to remove dampness,to eliminate stasis to subdue tumefaction.

Indications: Venomous snake bites.

Usage and Dosage: Oral administration: make decoction, 12 to 20 g. External use: smash and apply.

● Passifloraceae

Passiflora foetida L.

Common name: Passion fruit

Property: Herbaceous liana.

Habitat Distribution: Grown on the grass slopes and roadsides at an altitude of 120 to 500 meters; native to the West Indies; distributed in pantropical regions.

Medicinal Part: Whole plant or fruit.

Efficacies: To clear lung-heat to relief cough, and to remove toxic substances to reduce swelling.

Indications: Often used for cough due to lung-heat, turbid urine ulcerative carbuncles, traumatic keratitis, lymphadenitis.

Usage and Dosage: Oral administration: make decoction, 9 to 15 g. External use:smash with appropriate amount of fresh leaves and apply.

● Lecythidaceae

Barringtonia racemosa (L.) Spreng.

Common name: Powder-puff tree

Property: Evergreen arbor.

Habitat Distribution: Grown in the forests of coastal regions; widely distributed in the tropical and subtropical regions of Africa, Asia and Oceania.

Medicinal Part: Root or fruit.

Efficacies: To purge pathogenic fire and clear away heat, and to relieve cough and asthma.

Indications: Root for bringing down heat; fruit for relieving cough.

Usage and Dosage: Oral administration: make decoction, 6 to 9 g.

● Melastomataceae

Melastoma dodecandrum Lour.

Property: Under-shrub.

Habitat Distribution: Grown in short grasses of hillsides; commonly found in acid soil; also distributed in China and Vietnam.

Medicinal Part: Whole plant or root.

Efficacies: To astringe the intestine to relieve dysentery, to promote blood circulation to relax muscles, to tonify blood to prevent miscarriage, to clear heat and remove dampness.

Indications: Smash and apply on sores, carbuncles, subcutaneous ulcers and furuncles; root for the treatment of cassava poisoning.

Usage and Dosage: Oral administration: make decoction, 15 to 30 g (the amount doubled for fresh herb); or crush to extract juice with fresh herb. External use: smash with appropriate amount and apply or wash with decoction.

Melastoma malabathricum L.

Common name: Singapore Rhododendron

Property: Shrub.

Habitat Distribution: Grown on hillsides, in wet or dry places, under valley or sparse forests, under thorny bamboo forests, or shrub tussock, on roadsides or beside ditches; distributed in China, the Indo-China Peninsula to Australia and Philippines.

Medicinal Part: Whole plant or root.

Efficacies: to eliminate stagnated food, to astringe to arrest bleeding, and to dissipate blood stasis to subdue swelling.

Indications: Whole herb for the treatment of indigestion, diarrhea anddysentery due to enteritis; smashing and applying or grinding into fine powder and sprinkling for the treatment of traumatic bleeding, knife and gunshot wounds; making decoction with roots for oral administration for hastening parturition by using pepper as a synergist , also known as oxytocic.

Usage and Dosage: Oral administration: make decoction, 15 to 30 g. External use: smash and apply or grind into fine powder and sprinkle.

● Combretaceae

Quisqualis indica L.

Common name: Rangoon creeper

Property: Climbing shrub.

Habitat Distribution: Distributed in China, India, Myanmar and Philippines; commonly cultivated.

Medicinal Part: Fruit, leaf and root.

Efficacies: To kill parasites, eliminated stagnated food , and to invigorate the spleen.

Indications: Fruit for the treatment of abdominal pain due ascariasis, infantile malnutritional stagnation, infantile dyspepsia, abdominal distension, diarrhea; leaf for the treatment of infantile malnutritional stagnation; root for the treatment of roundworms.

Usage and Dosage: Oral administration: smash with 9 to 12 g of fruits and make decoction; use 6 to 9 g of kernel in pills or powder or take alone and finish once or twice.

Terminalia catappa L.

Common name: Indian almond

Property: Megaphanerophyte.

Habitat Distribution: Grown on hot and humid coastal beaches; distributed in China, Malaysia, Vietnam, India and Oceania; also commonly seen in the tropical coasts of South America.

Medicinal Part: Seed or bark.

Efficacies: To clear heat and remove toxins.

Indications: Pharyngalgia, dysentery and pyogenic infections; bark for the treatment of gastric and bilious heat , diarrhea and dysentery.

Usage and Dosage: Oral administration: make decoction, 3 to 10 g.

Rhizophoraceae

Bruguiera gymnorrhiza (L.) Poir.

Common name: Large-leafed mangrove

Property: Shrub or arbor.

Habitat Distribution: Grown in coastal mudflats; distributed in southeast Africa, India, Sri Lanka, Malaysia, Thailand, Vietnam, Northern Australia and Polynesia.

Medicinal Part: Bark.

Efficacies: To astringe to relieve diarrhea.

Indications: Diarrhea, deficiency of spleen and kidney.

Usage and Dosage: Oral administration: make decoction, 3 to 9 g.

● Clusiaceae

Calophyllum inophyllum L.

Common name: Alexandrian laurel

Property: Arbor.

Habitat Distribution: Grown in open hilly spaces and coastal sand wastelands; distributed in China, India, Sri Lanka, the Indo-China Peninsula, Malaysia, Indonesia (Sumatra), Andaman Islands, the Philippines, Polynesia and Madagascar and Australia.

Medicinal Part: Root and leaf.

Efficacies: To remove blood stasis to relieve pain.

Indications: Rheumatic pain, traumatic injury, dysmenorrhea, traumatic bleeding.

Usage and Dosage: Oral administration: make decoction, 3 to 10 g. External use: smash with appropriate amount of fresh leaves and apply.

● Malvaceae

Abelmoschus moschatus Medicus

Common name: Musk okra, musk mellow, tropical jewel-hibiscus

Property: Annual or biennial herb.

Habitat Distribution: Often grown in plains, valleys, hillside shrubs or beside mountain streams; distributed in China, Vietnam, Laos, Cambodia, Thailand and India; widely cultivated in tropical regions.

Medicinal Part: Root, leaf and flower.

Efficacies: To clear heat and eliminate dampness, and to remove toxins and expel pus.

Indications: Root for the treatment of protracted hyperpyrexia, cough due to lung-heat, post-partum galactostasis constipation, amoebic dysentery, lithangiuria; leaf for the treatment of ulcerative carbuncles, panaris and fractures (external use); flower for the treatment of burns and scalds (external use).

Usage and Dosage: Oral administration: make decoction, 9 to 15 g. External use: smash with appropriate amount of fresh herb and apply.

Hibiscus tiliaceus Linn.

Common name: Sea hibiscus

Property: Evergreen shrub or arbor.

Habitat Distribution: Grown in coastal zones; distributed in tropical countries such as Vietnam, Cambodia, Laos, Myanmar, India, Indonesia, Malaysia and Philippines.

Medicinal Part: Leaf, bark or flower.

Efficacies: To clear lung-heat to relieve cough, to remonve removing toxic substances and subside swelling.

Indications: Cough due to lung-heat, swelling and pain of sores and furuncles, cassava poisoning.

Usage and Dosage: Oral administration: make decoction, 30 to 60 g; or crush to extract juice. External use: smash with appropriate amount and apply.

Sida acuta Burm. f.

Common name: Common wireweed

Property: Upright shrub-like herb.

Habitat Distribution: Often grown in hillside shrubs, on roadsides or barren slopes; native to India; distributed in Vietnam, China, and Laos.

Medicinal Part: Root and leaf.

Efficacies: To clear damp-heat, to remove toxicity and subside swelling, to promote blood circulation to relieve pain.

Indications: Damp-heat diarrhea, acute mastitis, hemorrhoids, ulcerative carbuncles traumatic injuries, fractures, traumatic bleeding.

Usage and Dosage: Oral administration: make decoction, 9 to 15 g. External use: wash with decoction of appropriate amount or smash with fresh herb and apply on the affected part.

Thespesia populnea (Linn.) Soland. ex Corr.

Common name: Bendy tree, potial tree, seaside mahoe

Property: Evergreen arbor.

Habitat Distribution: Grown in sunny places in the seaside and coasts; distributed in Vietnam, Cambodia, Sri Lanka, India, Thailand, the Philippines and tropical Africa.

Medicinal Part: Bark or leaf.

Efficacies: To relieve inflammation.

Indications: Bark for the treatment of dysentery, hemorrhoids and various skin diseases; leaf for relieving inflammation, subsideing swelling, and for headache, scabies and tinea.

Usage and Dosage: External use: wash with decoction of appropriate amount, or smash with fresh herb and apply on the affected part.

Urena lobata Linn.

Common name: Caesarweed

Property: Upright shrub-like herb.

Habitat Distribution: Grown in dry hot open area, on grass slopes or under sparse forests; distributed in China, Vietnam, Cambodia, Laos, Thailand, Myanmar, India and Japan.

Medicinal Part: Root and whole herb.

Efficacies: To dispel wind and invigoratie blood circulation , to clear heat and eliminate dampness, and to remove toxic substances to sugside swelling.

Indications: Root for the treatment of rheumatic arthralgia, cold, malaria, enteritis, dysentery, infantile dyspepsia and leucorrhea; whole herb for the treatment of traumatic injuries, fractures, venomous snake bites and mastitis (external use).

Usage and Dosage: Oral administration: make decoction, 30 to 60 g; or crush to extract juice; or infuse in wine. External use: smash and apply.

● Euphorbiaceae

Excoecaria agallocha Linn.

Common name: Blinding tree

Property: Evergreen arbor.

Habitat Distribution: Grown in humid coastal regions; distributed in China, India, Sri Lanka, Thailand, Cambodia, Vietnam, the Philippines and Oceania.

Medicinal Part: Whole plant.

Efficacies: To cause purgation and to eliminate toxic substances.

Indications: Constipation due to excessive heat, intractable skin ulcers, pyogenic infections of hand and foot.

Usage and Dosage: External use: smash and apply.

Euphorbia antiquorum L.

Common name: Indian spurge tree

Property: Fleshy shrub-like dungarunga.

Habitat Distribution: Native to India; cultivated in Micronesia; distributed in tropical Asia.

Medicinal Part: Whole plant.

Efficacies: To dissipate blood stasis to eliminate inflammation, and to clear heat and remove toxicity.

Indications: Stem and leaf with action for subsiding swelling, draw out toxin and relieve diarrhea, for the treatment of acute gastroenteritis, malaria, traumatic injuries; juice with action for causing drastic purgation, expelling retained water, and reliving itching for the treatment of cirrhotic ascites and tinea.

Usage and Dosage: Oral administration: peel 30 to 160 g of fresh stem and chop it into pieces, remove its juice, add 15 g of rice, fry to brown sallow, and then make decoction with two bowls of water.

Euphorbia cyathophora Murr.

Common name: Fire on the mountain

Property: Annual or perennial herb.

Habitat Distribution: Grown in roadside grasses; native to Central and South America; domesticated in Africa, the Old World.

Medicinal Part: Whole plant.

Efficacies: To arrest bleeding by regulating meridians, to rejoin the fractured bones, to subside swelling and to relieve cough.

Indications: Chronic bronchitis, traumatic injuries, traumatic bleeding, fractures.

Usage and Dosage: Oral administration: make decoction, 3 to 9 g. External use: smash with appropriate amount of fresh herb and apply.

Euphorbia heterophylla L.

Common name: Fireplant, painted euphorbia, Japanese poinsettia, desert poinsettia, wild poinsettia, fire on the mountain, paintedleaf, painted spurge, milkweed

Property: Perennial herb.

Habitat Distribution: Grown in hillside grasses and roadsides; native to North America; cultivated and domesticated in Africa, the Old World.

Medicinal Part: Whole herb.

Efficacies: To regulate menstruation, to arrest bleeding, to relieve cough, to rejoin the fractured bone, to disperse swelling.

Indications: Menorrhagia, traumatic injuries, fractures, coughs.

Usage and Dosage: Oral administration: make decoction, 3 to 9 g. External use: smash with appropriate amount of fresh herb and apply.

Euphorbia hirta L.

Common name: Garden Euphorbia

Property: Annual herb.

Habitat Distribution: Grown on sunny slopes, roadsides, in valleys or shrubs; distributed in the tropical and subtropical regions of the world.

Medicinal Part: Whole herb.

Efficacies: To clear heat and remove toxicity, to relieve itching through eliminating dampness, and to promote lactation.

Indications: Pulmonary abscess, acute mastitis, dysentery, diarrhea, heat strangury, hematuria, eczema, tinea pedis, cutaneous pruritus, furuncles, carbuncles, sores, ulcerative gingivitis, postpartum hypogalactia; fresh juice for the treatment of tinea (external use).

Usage and Dosage: Oral administration: 15 to 30 g. External use: smash with appropriate amount of fresh herb and apply on the affected part or wash with decoction.

Euphorbia prostrata Ait.

Common name: Prostrate Euphorbia

Property: Annual herb.

Habitat Distribution: Grown on roadsides, cottage surroundingsand wasteland shrubs; native to tropical and subtropical America; naturalized to the tropical and subtropical regions of the Old World.

Medicinal Part: Whole herb.

Efficacies: To clear heat and eliminate dampness, to cool the blood and to remove toxic materials.

Indications: Lactagogue, dysentery, diarrhea, diphtheria, hypogalactia, gringival hemorrhage, hemafecia, gonorrhea, hematuria, herpes zostor, carbuncles and furuncles, eczema.

Usage and Dosage: Oral administration: make decoction (fresh herb 30 to 60 g); or crush to extract juice. External use: smash with appropriate amount of fresh herb and apply.

Euphorbia thymifolia L.

Common name: Tymifolious Euphorbia

Property: Annual herb.

Habitat Distribution: Grown on roadsides, cottage surroundings, grasses, sparse shrubs; commonly seen in sandy soil; widely distributed in the tropical and subtropical regions of the world (except Australia).

Medicinal Part: Whole herb.

Efficacies: To clear heat to remove dampness, to give astringents and to relieve itching.

Indications: Bacillary dysentery, enteritis, diarrhea.

Usage and Dosage: Oral administration: make decoction, 15 to 30 g (fresh herb 30 to 60 g); or crush to extract juice for decoction. External use: smash and apply or wash with decoction.

Macaranga carolinensis Volkens

Common name: Caroline macaranga

Property: Shrub or arbor.

Habitat Distribution: Grown in sparse forests of wildernesses and on sunny slopes; distributed in the Pacific Sulawesi Island, the Caroline Islands and the Gilbert Islands.

Medicinal Part: Twigs, leaf and root.

Efficacies: to relieve inflammation, and to remove toxic substance.

Indications: Gynecopathy, diabetes, infantile crying and anorexia, swollen sores.

Usage and Dosage: Oral administration: make decoction, 3 to 9 g.

Phyllanthus amarus Shumacher et Thonning

Common name: Leaf flower

Property: Annual herb.

Habitat Distribution: Grown in wild grasslands, sunny valleys or on hillsides; distributed in China, India, the Indo-China Peninsula, Malaysia, Philippines to the tropical America.

Medicinal Part: Whole herb.

Efficacies: To invigorate the spleen to relieve dyspepsia, to promote diuresis to relieve stranguria, to clear toxic-heat.

Indications: Infantile malnutrition, dysentery, gonorrhea, acute mastitis, ulcerative gingivitis, venomous snake bites.

Usage and Dosage: Oral administration: make decoction, 15 to 30 g (fresh herb 1,500 to 3,000 g); or crush to extract juice. External use: smash and apply.

Phyllanthus urinaria L.

Common name: Chamber bitter, gripeweed, shatterstone, stonebreaker or leafflower.

Property: Annual herb.

Habitat Distribution: Grown in wildernesses, dry fields, on mountainous roadsides or forest edges at an altitude below 500 meters; distributed in China, India, Sri Lanka, the Indo-China Peninsula, Japan, Malaysia, Indonesia to South America.

Medicinal Part: Whole herb.

Efficacies: To remove toxic substances, to relieve inflammation, to clear heat and relieve diarrhea, and to promote diuresis.

Indications: Painful conjunctival congestion, enteritic diarrhea, dysentery, hepatitis, infantile malnutrition, nephritic edema and urethritis.

Usage and Dosage: Oral administration: make decoction, 15 to 30 g. External use: smash with appropriate amount of fresh herb and apply around the wound.

● Fabaceae

Tamarindus indica Linn.

Common name: Tamarind

Property: Arbor.

Habitat Distribution: Native to Africa; cultivated in tropical regions.

Medicinal Part: Fruit.

Efficacies: To cause purgation to cool the bowels, to dispel wind and counter scurvy.

Indications: Diarrhea, flatulence, leprosy, paralysis, paraplegia, scurvy, hypercholia, human parasites, relieving alcoholism and daturism.

Usage and Dosage: Oral administration: make condensed decoction with 15 to 30 g of fruit and white sugar, respectively, and take twice daily, in the morning and evening respectively.

Abrus precatorius Linn.

Common name: Coral-head plant, Love pea

Property: Liana.

Habitat Distribution: Grown on hilly areas, or in intermontane and roadside shrubs; widely distributed in tropical regions.

Medicinal Part: Stem leaf and root.

Efficacies: To clear heat and remove toxicity, to promote diuresis, to induce emesis, to expel parasites, to remove toxic substances, and reduce swelling, to promote salivation , to moister the lung, to clear heat, and to promote diuresis.

Indications: , Scabies and tinea, carbuncles, sore throat, hepatitis, cough and asthma due to excessive phlegm.

Usage and Dosage: Oral administration: make decoction, 9 to 15 g. External use: wash with decoction or smash and apply.

Note: The seeds are highly toxic.

Derris trifoliata Lour.

Common name: Trifoliate Jewelvine, Marshy Jewelvine

Property: Climbing shrub.

Habitat Distribution: Often grown in coastal shrubs, seaside bushes or mangroves near the shore ; distributed in China, India, Malaysia and northern Australia.

Medicinal Part: Twigs and leaf.

Efficacies: To dissipate stasis to relieve pain, and to kill parasites.

Indications: Eczema, rheumatic arthritis, traumatic injury (without unbroken skin) (external use).

Usage and Dosage: External use: wash with decoction of appropriate amount, or grind into fine powder and apply.

Moraceae

Cannabis sativa L.

Common name: Hemp fimble

Property: Upright annual herb.

Habitat Distribution: Native to Sikkim, Bhutan, India, China and Central Asia; cultivated in Micronesia.

Medicinal Part: Flower, fruit and leaf.

Efficacies: To dispel pathogenic wind, Dispelling wind, to remove dampness, to promote the circulation of Qi, and to remove stasis.

Indications: Rheumatic arthralgia, stuffiness of chest and diaphragm, malaria, dysentery, diarrhea, postpartum cold, hepatitis, hemorrhoids, acariasis; leaves containing anesthetic resin can be used to formulate anesthetics.

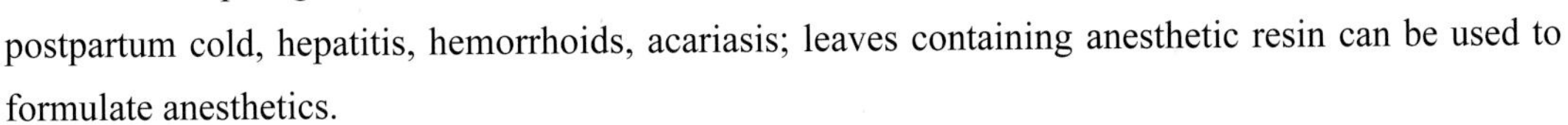

Usage and Dosage: Oral administration: make decoction, 9 to 18g; or use in pills or powder. External use: smash and apply or extract oil and apply.

● Urticaceae

Pilea microphylla (L.) Liebm.

Common name: Rockweed, Artillery plant, Gunpowder plant or Brilhantina

Property: Slender small herb.

Habitat Distribution: Often grown in roadside stone cracks and sheltered places on walls; native to tropical South America and later introduced to tropical Asia and Africa.

Medicinal Part: Whole herb.

Efficacies: To clear heat and remove toxicity.

Indications: Sores and carbuncles, undefined swelling and sore; external use for scald and burns.

Usage and Dosage: Oral administration: make decoction, 5 to 15 g. External use: smash with appropriate amount of fresh whole herb and apply; or squeeze juice and apply.

Pouzolzia zeylanica (L.) Benn.

Common name: Pouzolz's bush

Property: Perennial herb.

Habitat Distribution: Grown on flat grasslands or near paddy fields, in shrubs of hills or low mountains, sparse forests or near ditches; widely distributed in tropical Asia.

Medicinal Part: Whole herb.

Efficacies: To clear heat and remove toxicity, to subside swelling and evacuate pus, to promote diuresis to relieve-stranguria.

Indications: Sores and carbuncles, acute mastitis, wind-fire toothache, dysentery, diarrhea, stranguria, gonorrhea.

Usage and Dosage: Oral administration: make decoction, 15 to 30 g (fresh whole herb 30 to 60 g). External use: smash and apply or crush to extract juice for mouthwash.

● Vitaceae

Cayratia trifolia (L.) Domin

Common name: Trifoliate cayratia

Property: Woody liana.

Habitat Distribution: Grown on hillsides, on forest edges near creeks or in forests; distributed in China, Vietnam, Laos, Cambodia, Thailand, Bangladesh, India, Malaysia and Indonesia.

Medicinal Part: Whole herb.

Efficacies: To alleviate inflammation to relieve pain, to dissipate blood stasis to activate blood circulation , to dispel wind-damp.

Indications: Traumatic injuries, fractures, rheumatic osteodynia , lumbar muscle strain, eczema, skin ulcer, pulmonary abscess, sores and furuncles.

Usage and Dosage: Oral administration: make decoction, 6 to 12g; also suitable for external use.

● Anacardiaceae

Spondias dulcis Parkinson

Common name: Golden Apple

Property: Deciduous arbor.

Habitat Distribution: Grown in shrubs or sparse forests of hills or low mountains or near houses; native to the Pacific islands such as Polynesia.

Medicinal Part: Fruit, leaf, bark and sap.

Efficacies: To alleviate and arrest bleeding.

Indications: Sore throat, stomatitis, ulcers, diarrhea, dysentery, eye inflammation, postpartum hemorrhage, cough, fever, stomachache.

Usage and Dosage: Oral administration: make decoction, 20 to 30 g. External use: 100 g.

● Apiaceae

Centella asiatica (L.) Urban

Common name: Asiatic pennywort

Property: Perennial herb.

Habitat Distribution: Grown on dark and humid grasslands or near ditches; distributed in China, India, Sri Lanka, Malaysia, Indonesia, Oceania Islands, Japan, Australia, Central Africa and South Africa.

Medicinal Part: Whole herb.

Efficacies: To clear heat and eliminate damp, to subside swelling and remove toxicity.

Indications: Eruptive abdominal disease, summer diarrhea, dysentery, Damp-heat jaundice, urethral calculus, stranguria complicated by hematuria, hematemesis, hemoptysis, conjunctival congestion, laryngeal swelling, rubella, acariasis, furuncles and carbuncles, traumatic injuries.

Usage and Dosage: Oral administration: make decoction, 9 to 15 g (fresh herb 15 to 30 g); or crush to extract juice. External use: smash and apply or crush to extract juice and apply.

● Apocynaceae

Catharanthus roseus (L.) G. Don

Common name: Madagascar Periwinkle

Property: Subshrub.

Habitat Distribution: Native to East Africa; cultivated in tropical and subtropical regions.

Medicinal Part: Whole plant.

Efficacies: To tranquilize and calm the mind, to pacify the liver and reduce blood pressure, and to prevent cancer.

Indications: Hypertension; eukemia, lymphoma, lung cancer, chorionic epithelioma, and metrocarcinoma due to the existence of vinblastine in the plants.

Usage and Dosage: Oral administration: make decoction, 6 to 15 g; or make extracts for injection.

Cerbera manghas Linn.

Common name: Common Cerberus Tree

Property: Arbor.

Habitat Distribution: Grown on seaside or in humid places near seaside; distributed in tropical Asia and Australia.

Medicinal Part: Seed.

Efficacies: To induce vomiting and purgation.

Indications: Bark, leaf and sap as pharmaceutical agents for inducing vomiting, purgation and abortion.

Usage and Dosage: The dosage should be cautious as excessive administration is lethal.

Note: The whole plant is toxic, and the fruit is highly toxic. A small amount can be lethal, and the toxicity is greater after roasting. The peel contains mangiferine, picrotoxin, cerberine, toxic quassin, cyanic acid, all with high toxicity.

Tabernaemontana divaricata (Linnaeus) R. Brown ex Roemer & Schultes

Common name: Crape jasmine, Pinwheel flower

Property: Shrub.

Habitat Distribution: Grown on forests edges; distributed in China and India; widely cultivated in tropical and subtropical Asia.

Medicinal Part: Leaf or root.

Efficacies: To reduce blood pressure, to clear heat by cooling, to promote diuresis to subside edema.

Indications: Oculopathy, scabies, breast sore, mad dog bites; root for the treatment of headaches and fractures.

Usage and Dosage: Oral administration: make decoction, 10 to 30 g. External use: smash with appropriate amount of fresh herb and apply.

Plumeria rubra L. 'Acutifolia'

Common name: Mexican frangipani

Property: Dungarunga.

Habitat Distribution: Native to South America; widely distributed in tropical and subtropical Asia.

Medicinal Part: Flower or bark.

Efficacies: Clearing heat, removing dampness, relieving summer-heat.

Indications: Cold fever, cough due to lung-heat, damp-heat jaundice, diarrhea, lithangiuria, heat stroke.

Usage and Dosage: Oral administration: make decoction, 4.5 to 12g.

● Asclepiadaceae

Calotropis procera (L.)Dry. ex Ait. f.

Common name: Rooster tree

Property: Upright shrub.

Habitat Distribution: Grown on low-altitude sunny slopes, seaside and open land; distributed in China, India, Sri Lanka, Myanmar, Vietnam and Malaysia.

Medicinal Part: Root, stem, leaf and fruit.

Efficacies: Diminishing inflammation, offering antibiosis , dissipating phlegm and removing toxicity.

Indications: Leprosy, asthma, cough, ulcers and tumors.

Usage and Dosage: Oral administration: make decoction, 1 to 3 g; or use in powder.

● Rubiaceae

Gardenia jasminoides Ellis

Common name: Cape Jasmine

Property: Shrub.

Habitat Distribution: Grown on hills, hillsides, in low-altitude wildernesses, valleys, forests or shrubs beside streams; distributed in China, Japan, North Korea, Vietnam, Laos, Cambodia, India, Nepal, Pakistan, Pacific Islands and North America.

Medicinal Part: Fruit.

Efficacies: Protecting liver, normalizing gallbladder function, reducing blood pressure, sedation, stopping bleeding, dispersing swelling.

Indications: Jaundice hepatitis, sprains and bruises, hypertension, diabetes.

Usage and Dosage: Oral administration: make decoction, 5 to 10 g. External use: grind into fine powder and apply.

Morinda citrifolia L.

Common name: Indian mulberry

Property: Shrub to dungarunga.

Habitat Distribution: Grown on coastal flat land or under sparse forests; distributed widely from India and Sri Lanka, the Indo-China Peninsula, south to North Australia, east to Polynesia and their islands.

Medicinal Part: Fruit.

Efficacies: Improving immunity.

Indications: Asthma and other respiratory diseases, diabetes, nephritis, arthritis, hysteria, allergies, arteriosclerosis, menstrual disorders, cardiovascular diseases (hypertension and myocardial infarction).

Usage and Dosage: Oral administration: extract enzymes or use fresh fruits, 60 to 120 g.

● Asteraceae

Bidens pilosa L. var. *radiata* Sch.-Bip.

Common name: Railway beggearticks

Property: Annual herb.

Habitat Distribution: Grown beside villages, on roadsides and in wastelands; widely distributed in tropical and subtropical Asia and America.

Medicinal Part: Whole herb.

Efficacies: Clearing heat and removing toxicity, dispersing stasis and invigorating the circulation of blood.

Indications: Upper respiratory tract infection, swollen sore throat, acute appendicitis, acute jaundice hepatitis, gastroenteritis, rheumatic arthralgia, malaria; sores and furuncles, venomous snake bites, traumatic injuries (external use).

Usage and Dosage: Oral administration: make decoction, 15 to 30 g.

Chromolaena odoratum (L.) R. King et H. Rob.

Common name: Siam grasss

Property: Perennial herb.

Habitat Distribution: Native to America; grown on low-altitude hilly lands, in shrubs and savannas.

Medicinal Part: Whole herb.

Efficacies: Killing parasites, stopping bleeding.

Indications: Traumatic injuries, sores, furuncles and carbuncles, paddy-field dermatitis, traumatic bleeding, uncontrollable bleeding after land leech bites.

Usage and Dosage: External use: smash with appropriate amount of fresh leaves and apply on the wound.

Crassocephalum crepidioides (Benth.) S. Moore

Common name: Okinawa Spinach

Property: Upright annual herb.

Habitat Distribution: Grown on hillsides, watersides and in shrubs; distributed in China, Thailand, Southeast Asia and Africa; a kind of weed widely distributed in pantropical regions.

Medicinal Part: Whole herb.

Efficacies: Tonifying spleen, dispersing swelling.

Indications: Indigestion, spleen deficiency and edema.

Usage and Dosage: Oral administration: make decoction, 15 to 30 g. External use: smash and apply.

Eclipta prostrata (L.) L.

Common name: Yerbadetajo

Property: Annual herb.

Habitat Distribution: Grown on riversides, roadsides or in paddy fields; widely distributed in tropical and subtropical regions of the world.

Medicinal Part: Whole herb.

Efficacies: Tonifying the liver and kidney, cooling blood to stop bleeding.

Indications: Hematemesis, epistaxis, hemoptysis, intestinal bleeding, hematuria, hemorrhoidal bleeding, metrorrhagia; additionally crushing to extract juice and applying on eyebrows and hair to promote their growth; oral administration for hair blacking.

Usage and Dosage: Oral administration: make decoction, 9 to 30 g; or make condensed decoction; or crush to extract juice; or use in pills or powder. External use: smash with appropriate amount and apply; pound to flosses and stuff into nose; or grind into fine powder and apply.

Elephantopus tomentosus L.

Common name: Tomentose elephantfoot

Property: Perennial herb.

Habitat Distribution: Grown in the wildernesses, shrubs of hillsides or on roadsides; widely distributed in tropical regions.

Medicinal Part: Whole herb.

Efficacies: Clearing heat and removing toxicity, cooling blood and inducing diuresis.

Indications: Epistaxis, jaundice, stranguria, dermatophytosis, edema, sores furuncles, snake and insect bites.

Usage and Dosage: Oral administration: make decoction, 25 to 50 g. External use: smash with appropriate amount of fresh herb and apply on the affected part.

Emilia sonchifolia (L.) DC.

Common name: Sowthistle taself flower, Red taself flower

Property: Annual herb.

Habitat Distribution: Grown in the wildernesses, shrubs of hillsides or on roadsides; widely distributed in tropical regions.

Medicinal Part: Whole herb.

Efficacies: Diminishing inflammation, relieving dysentery.

Indications: Parotitis, mastitis, infantile malnutritional stagnation, skin eczema.

Usage and Dosage: Oral administration: make decoction, 9 to 18g (fresh herb 15 to 30 g); or crush to extract juice and take it in pharynx. External use: wash with decoction of appropriate amount; or smash and apply.

Glossogyne tenuifolia Cass.

Common name: Thinleaf glossogyne

Property: Perennial herb.

Habitat Distribution: Grown on hard sands, open sands and seaside; distributed in China, the Philippines, Malaysia and Oceania.

Medicinal Part: Whole herb.

Efficacies: Clearing heat and removing toxicity, invigorating the circulation of blood and removing stasis.

Indications: Cold fever, swollen sore throat, enteritis diarrhea, appendicitis, traumatic injuries, ulcers and furuncles, subcutaneous ulcers of gonorrhea, toothache, zonal eczema.

Usage and Dosage: Oral administration: make decoction, 25 to 50 g. External use: smash and apply.

Gynura procumbens (Lour.) Merr.

Common name: Climbing velvetplant

Property: Climbing herb.

Habitat Distribution: Grown on sandy soil of the slopes beside streams within forests or, climbing on shrubs or arbors; distributed in China, Vietnam, Thailand, Indonesia and Africa.

Medicinal Part: Branch and leaf.

Efficacies: Promoting menstruation and activating collaterals, diminishing inflammation and relieving cough, removing stasis and dispersing swelling, invigorating the circulation of blood and promoting granulation.

Indications: Traumatic injuries, rheumatic arthralgia and gout.

Usage and Dosage: Oral administration: make decoction, 150 to 250 g; take as vegetables.

Synedrella nodiflora (L.) Gaertn.

Common name: Nodalflower, sysnedrella

Property: Annual herb.

Habitat Distribution: Grown in wildernesses, on cultivated lands, roadsides and house surroundings, with strong fecundity; native to America; now widely distributed in tropical and subtropical regions of the world.

Medicinal Part: Whole herb.

Efficacies: Clearing heat and promoting eruption, removing toxicity and dispersing swelling.

Indications: Cold fever, ecchymosis, ulcerative carbuncles.

Usage and Dosage: Oral administration: make decoction, 15 to 30 g. External use: smash with appropriate amount and apply; or wash with decoction.

Vernonia cinerea (L.) Less.

Common name: Ashycoloured ironweed

Property: Annual or perennial herb.

Habitat Distribution: Often grown in hillside open land , on wastelands, field sides and roadsides; distributed in India to the Indo-China Peninsula, Japan, Indonesia and Africa.

Medicinal Part: Whole herb.

Efficacies: Dispelling wind and dissipating heat, drawing out toxin and dispersing swelling, tranquilizing the mind and calming the nerves, promoting digestion to eliminate stagnation.

Indications: Cold, fever, neurasthenia, insomnia, dysentery, traumatic injuries, snake injuries, mastitis, sores and furuncles.

Usage and Dosage: External use: smash with appropriate amount of fresh herb and apply on the affected part (dried herb 15 to 30 g, fresh herb 30 to 60 g).

Wedelia biflora (Linn.) DC.

Common name: Sea Daisy, Sea Ox-eye

Property: Climbing herb.

Habitat Distribution: Grown on grasslands, in shrubs or under forests and also often seen in coastal dry sands; distributed in India, the Indo-China Peninsula, Indonesia, Malaysia, the Philippines, Japan and Oceania.

Medicinal Part: Leaf and flower.

Efficacies: Relieving pain and diarrhea.

Indications: Acnes, diarrhea, stomachache.

Usage and Dosage: Oral administration: make decoction, 15 to 30 g (fresh herb 30 to 60 g). External use: smash with appropriate amount and apply; or crush to extract juice for mouthwash.

Campanulaceae

Hippobroma longiflora (Linnaeus) G. Don Gen.

Common name: Star of Bethlehem

Property: Perennial upright herb.

Habitat Distribution: Grown on forest edges and roadsides; distributed in tropical America, Oceania, and the West Indies.

Medicinal Part: Whole herb.

Efficacies: Removing toxicity, dispersing swelling, relieving pain.

Indications: Asthma.

Usage and Dosage: Oral administration: make decoction, 0.5 to 3g. The dosage should not be excessive, so as to avoid intoxication accident.

● Boraginaceae

Messerschmidia argentea (L. f.) Johnst.

Common name: Silverhair messerschmidia

Property: Dungarunga or shrub.

Habitat Distribution: Grown on coastal sands; distributed in China, Japan, Vietnam and Sri Lanka.

Medicinal Part: Branch and leaf.

Efficacies: Removing toxicity, diminishing inflammation.

Indications: Marine thorny creatures poisoning; leaves containing rosmarinic acid and derivatives used for anti-viruses, anti-bacteria, anti-oxidantion and diminishing inflammation.

Usage and Dosage: Oral administration: make decoction, 30 to 60 g; or use in powder. External use: make condensed decoction and apply.

● Solanaceae

Physalis angulata L.

Common name: Cutleaf groundcherry

Property: Perennial herb.

Habitat Distribution: Grown under valley forests, around villages or along roadsides at low altitude location; Native to China, Japan, India, Australia and America.

Medicinal Part: Whole herb.

Efficacies: Clearing heat and removing toxicity, reducing swelling and promoting diuresis.

Indications: Swollen sore throat, parotitis, acute and chronic bronchitis, pulmonary abscess, dysentery, orchitis, dysuria; external sue for pemphigus.

Usage and Dosage: Oral administration: make decoction, 15 to 30 g; or juice extraction before administration, For external use at proper amount, smash and apply, or make decoction for gargling or for fumigation.

Physalis peruviana L.

Common name: Peru groundecherry

Property: Perennial herb.

Habitat Distribution: Grown on roadsides or in river valleys; Native to South America.

Medicinal Part: Whole herb.

Efficacies: Clearing heat and removing toxicity, promoting diuresis.

Indications: Fever due to eruptive diseases, cold, parotiditis , swollen sore throat, cough, orchitis, pemphigus.

Usage and Dosage: Oral administration: make decoction, 15 to 30 g.

Solanum americanum Mill.

Common name: American Nightshade

Property: Delicate herb.

Habitat Distribution: Grown on riversides, in moist places of jungles or wastelands on forest edges; distributed in the Malay Islands.

Medicinal Part: Leaf.

Efficacies: Clearing heat and promoting diuresis, cooling blood and relieving toxin, diminishing inflammation and swelling.

Indications: Dysentery, hypertension, jaundice, tonsillitis, cough due to lung-heat, gingival bleeding; noxious damp of skin, skin blisters and rat bites (external use).

Usage and Dosage: Oral administration: make decoction, 9 to 15 g, or crush to extract juice. External use: smash with appropriate amount and apply.

Convolvulaceae

Ipomoea digitata L.

Common name: Giant potato

Property: Large perennial winding herb.

Habitat Distribution: Grown in seaside coppices, in sparse mountain forests or shrubs beside creeks; distributed in tropical regions such as China and Vietnam.

Medicinal Part: Root or leaf.

Efficacies: Removing toxicity and resolving mass, purging retained fluid to subside swelling, replenishing blood.

Indications: Edema and abdominal distension, constipation, scrofula , carbuncles.

Usage and Dosage: Oral administration: make decoction, 5 to 10 g, or crush to extract juice. External use: appropriate amount.

Ipomoea triloba L.

Common name: Ivy-like Merrenmia

Property: Twining or creeping herb.

Habitat Distribution: Grown in shrubs or roadside grasses; distributed in tropical Africa, Mascarene Islands, tropical Asia from India, Sri Lanka, east to Myanmar, Thailand, Vietnam, entire Malaysia, Caroline Islands to Queensland, Australia; also seen in the Christmas Island in the middle of the Pacific Ocean.

Medicinal Part: Whole herb.

Efficacies: Clearing heat and removing toxicity, relieving sore throat.

Indications: Cold, acute tonsillitis, pharyngitis, acute conjunctivitis.

Usage and Dosage: Oral administration: make decoction, 15 to 30 g.

● Scrophulariaceae

Lindernia antipoda (L.) Alston

Common name: Creeping falsepimpernel

Property: Annual herb.

Habitat Distribution: Often grown beside paddy fields and humid grasslands; widely distributed from India to the tropical and subtropical regions from India to northern Australia.

Medicinal Part: Whole herb.

Efficacies: Eliminating stasis, dispersing swelling, promoting diuresis, removing toxicity.

Indications: Traumatic injuries, swelling ulcers and furuncles, gonorrhea.

Usage and Dosage: Crush to extract juice with 90 to 120 g of fresh leaves, and infuse in wine before taking; apply the dregs on the affected part.

Lindernia crustacea (L.) F. Muell

Common name: Brittle falsepimpernel

Property: Herb.

Habitat Distribution: Grown in low-humid places such as fields, grasslands and roadsides; widely distributed in tropical and subtropical regions.

Efficacies: Clearing heat and eliminating damp, invigorating the circulation of blood and relieving pain.

Indications: Wind-heat common Anemopyretic cold, damp-heat diarrhea, nephritis edema, leucorrhea, irregular menstruation, sores, carbuncles and furuncles, venomous snake bites, traumatic injuries.

Usage and Dosage: Oral administration: make decoction, 10 to 15 g (fresh herb 30 to 60 g); or grind into fine powder and infuse in wine.

● Acanthaceae

Asystasia gangetica (L.) T. Anders.

Common name: Ganges Asystasia

Property: Perennial herb.

Habitat Distribution: A kind of pantropical weed; grown on forest edges and roadsides; distributed in China, India, Thailand, the Indo-China Peninsula to Malaya.

Medicinal Part: Whole herb.

Efficacies: Promoting reunion of fractured bones, removing toxicity and relieving pain, cooling blood for hemostasis.

Indications: Traumatic bone fractures, ecchymoma; a key medicine for traumas; both oral administration and external application have certain effects on sores, carbuncles and furuncles as well as venomous snake bites, in which fresh herbs are better; in the treatment of various hemorrhage syndromes caused by blood-heat, it has a special feature of arresting blooding without ecchymosis, especially effective for those hemostasis with ecchymosis; frequently used in traumatic bleeding.

Usage and Dosage: Oral administration: make decoction, 9 to 15 g. External use: smash with appropriate amount and apply on the affected part or grind into fine powder.

Ruellia repens Linnaeus

Common name: Creeping ruellia

Property: Perennial spreading herb.

Habitat Distribution: Grown on low-altitude roadsides or open land and grasslands; distributed in China, India, Malaysia and the Philippines.

Medicinal Part: Leaf.

Efficacies: Removing toxicity, dispersing swelling, relieving pain.

Indications: Subaxillary abscess, knife incised wounds, abdominal pain, toothache.

Usage and Dosage: Oral administration:30 to 50 g. External use: appropriate amount.

● Verbenaceae

Callicarpa candicans (Burm. f.) Hochr. var. *ponapensis* Fosberg

Common name: Whitehairy beautyberry

Property: Shrub.

Habitat Distribution: Grown on plains, hillsides, roadsides or open and deserted places; distributed in Pohnpei, Micronesia and Kosrae.

Medicinal Part: Root, bark and leaf.

Efficacies: Dispelling dampness and relieving itching, killing parasites.

Indications: Pruritus cutanea caused by scabies, eczemas and neurodermatitis.

Usage and Dosage: External use: make decoction with roots, barks and leaves for washing. This herb, which is toxic, is forbidden for oral administration.

Clerodendrum inerme (L.) Gaertn.

Common name: Unarmed glorybower

Property: Climbing shrub.

Habitat Distribution: Grown on coastal beaches and the places where tides can reach; distributed in South China, India, Southeast Asia and the north of Oceania.

Medicinal Part: Root.

Efficacies: Removing stasis, dispersing swelling, removing dampness, destroying parasites.

Indications: Traumatic ecchymoma, skin eczema, scabies.

Usage and Dosage: Oral administration: crush to extract juice and drink. External use: make decoction for fumigation and washing; smash and apply or grind into fine powder and Spread.

Phyla nodiflora (L.) Greene

Common name: Knottedflower phyla

Property: Perennial herb.

Habitat Distribution: Grown in wet places on hillsides, flatlands and bench lands; distributed in the tropical and subtropical regions of the world.

Medicinal Part: Whole herb.

Efficacies: Dispersing blood stasis and promoting tissue regeneration, and promoting urination.

Indications: Cough, hematemesis, stranguria, dysentery, toothache, furuncles, occipitalgia, shingles and traumatic injuries.

Usage and Dosage: Oral administration: make decoction, 15 to 30 g. External use: smash with appropriate amount of fresh herb and apply on the affected part.

Stachytarpheta jamaicensis (L.) Vahl.

Common name: Jamaica falseyalerian

Property: Perennial sturdy herb or subshrub.

Habitat Distribution: Grown in the grasses of dark and moist places of valleys at an altitude from 300 to 580 meters; native to Central and South America; widely distributed in Southeast Asia.

Medicinal Part: Whole herb.

Efficacies: Clearing heat and removing toxicity, inducing diuresis and treating stranguria.

Indications: Lithangiuria, urinary tract infections, rheumatic bone pain, laryngitis, acute conjunctivitis, swelling and pain of carbuncles and furuncles.

Usage and Dosage: Oral administration: make decoction, 9 to 15 g; or infuse in wine. External use: smash with appropriate amount and apply.

Vitex negundo L.

Common name: Negundo chastetree

Property: Shrub or dungarunga.

Habitat Distribution: Grown on roadsides on hillsides or in shrubs; distributed in East Africa, Madagascar, Southeast Asia and Bolivia of South America.

Medicinal Part: Root or seed.

Efficacies: Dispelling wind to relieve exterior syndrome, relieving cough and asthma, regulating Qi-flowing to promote digestion and relieve pain.

Indications: Cold, cough, asthma, stomachache, acid regurgitation, indigestion, food retention and diarrhea, cholecystitis, gallstones, hernias; root for expelling pinworms.

Usage and Dosage: Oral administration: make decoction, 15 g. External use: appropriate amount.

Lamiaceae

Clerodendranthus spicatus (Thunb.) C. Y. Wu

Common name: Spicate clerodendranthus

Property: Perennial herb.

Habitat Distribution: Grown in the wet places under forests and sometimes also seen on unshaded flats; mostly cultivated; distributed in China, India, Myanmar, Thailand, Indonesia, the Philippines, Australia and neighboring islands.

Medicinal Part: Aerial part.

Efficacies: Clearing heat and dispelling dampness, promoting diuresis for discharging urinary calculus.

Indications: Acute and chronic nephritis, cystitis, lithangiuria and rheumatic arthritis, especially effective to kidney disease.

Usage and Dosage: Oral administration: make decoction, 30 to 60 g (fresh herb 90 to 120 g).

Hyptis rhomboidea Mart. et Gal.

Common name: rhomboid bushmint

Property: Annual herb.

Habitat Distribution: Grown on open wastelands; native to tropical America; now widely distributed in all tropical regions.

Medicinal Part: Whole herb.

Efficacies: Dispelling dampness,promoting digestion to eliminate stagnation dispersing swelling, relieving fever, stopping bleeding.

Indications: Cold, lung disease, heat stroke, asthma, gonorrhea.

Usage and Dosage: Oral administration: make decoction, 30 to 60 g.

Ocimum sanctum L.

Common name: Skulleaplike coleus

Property: Subshrub herb.

Habitat Distribution: A kind of pantropical weed; grown on dry sandy grassland; distributed from North Africa, West Asia, India, the Indo-China Peninsula, Malaysia, Indonesia, Philippines to Australia.

Medicinal Part: Whole herb or leaf.

Efficacies: Relieving pain and asthma.

Indications: Headache, cold, stomachache, inflammation, heart disease, toxicosis, malaria, asthma.

Usage and Dosage: Oral administration: make decoction, 3 to 9 g; or grind into fine powder.

● Commelinaceae

Commelina diffusa Burm. f.

Common name: Climbing dayflower

Property: Annual spreading herb.

Habitat Distribution: Grown under forests, in shrubs, beside creeks, or in wet open land; widely distributed in the tropical and subtropical regions of the world.

Medicinal Part: Whole herb.

Efficacies: Clearing heat and removing toxicity, promoting diuresis and dispersing swelling, stopping bleeding.

Indications: Acute pharyngitis, dysentery, sores and furuncles, dysuria; traumatic bleeding (external use).

Usage and Dosage: Oral administration: make decoction (fresh herb 30 to 60 g). External use: grind into fine powder and apply on the wound.

● Flagellariaceae

Flagellaria indica Linn.

Common name: Indian flagellaria

Property: Perennial climbing herb.

Habitat Distribution: Grown on ditch edges and sparse riverside forests in the coastal regions at an altitude from 40 to 450 meters; distributed in China, India, the Indo-China Peninsula, the Philippines, Indonesia and Australia.

Medicinal Part: Stem and rhizome.

Efficacies: Promoting diuresis.

Indications: Edema and dysuria.

Usage and Dosage: Oral administration: make decoction, 9 to 15 g.

Zingiberaceae

Costus speciosus (J. König) Smith

Common name: Crepe ginger

Property: Perennial herb.

Habitat Distribution: Grown under sparse forests, on shade valley wetlands, roadside grasses, barren slopes and ditch edges; widely distributed in tropical Asia.

Medicinal Part: Rhizome.

Efficacies: Diminishing inflammation and promoting diuresis, dissipating blood stasis to subside swelling, removing toxicity and relieving itching.

Indications: Pertussis, nephritic edema, urinary tract infection, cirrhostic ascites, dysuria; urticaria, sores and furuncles, otitis media (external use).

Usage and Dosage: Oral administration: make decoction, 6 to 15 g. External use: wash with decoction of appropriate amount, or smash with appropriate amount of fresh herb and apply on the affected part.

Curcuma australasica Hook. f.

Common name: Cape York Lily

Property: Herb.

Habitat Distribution: Grown under forests, in shrubs or on village roadsides; distributed in Papua New Guinea and Australia; also cultivated in China.

Medicinal Part: Rhizome.

Efficacies: Resisting oxidation and preventing tumors, reducing blood fat and glucose, resisting ulceration, protecting liver, resisting myocardial ischemia, relieving depression, dispelling bacteria, diminishing inflammation, and dispelling virus and fungus.

Indications: Cancer and chronic diseases such as diabetes, coronary heart disease, arthritis, Alzheimer's disease.

Usage and Dosage: Oral administration: make decoction, 3 to 10 g, or use in pills or powder.

Curcuma longa L.

Common name: Common turmeric

Property: Perennial herb.

Habitat Distribution: Grown in sunny places; widely cultivated in East and Southeast Asia.

Medicinal Part: Rhizome.

Efficacies: Breaking blood stasis, promoting the circulation of Qi, promoting menstruation, and relieving pain.

Indications: Distending pain in chest and abdomen, arm pain, abdominal mass, Amenorrhea due to blood stasis, postpartum abdominal pain due to lochiostasis , traumatic injuries, carbuncles.

Usage and Dosage: Oral administration: make decoction, 10 to 20 g, or use in pills or powder.

Zingiber zerumbet (L.) Smith

Commo name: Zirumbet ginger

Property: Perennial herb.

Habitat Distribution: Grown in the shady wet places under forests; widely distributed in tropical Asia.

Medicinal Part: Rhizome.

Efficacies: Dispelling wind and removing toxicity.

Indications: Abdominal pain and diarrhea. Aromatic oil can be extracted as raw materials for blending essence. Tender shoots and leaves can be used as vegetables.

Usage and Dosage: Oral administration: make decoction, 9 to 15 g.

● Liliaceae

Aloe vera (Linnaeus) N. L. Burman

Common name: Really aloe

Property: Herb.

Habitat Distribution: Native to the tropical arid regions of Africa; distributed almost all over the world; Grown in wild in India, Malaysia, the African continent and even the whole tropical regions.

Medicinal Part: Leaf.

Efficacies: Purging fire to remove toxicity, dispersing stasis, destroying parasites.

Indications: Conjuntival congestion, constipation, gonorrhea, hematuria, infantile fright epilepsy, malnutritional stagnation, burns and scalds, amenorrhea, hemorrhoids, scabies, pyogenic infections of carbuncles and furuncles, traumatic injuries.

Usage and Dosage: Oral administration: use in pills or powder, or grind into fine powder and use in capsules, 0.6–1.5 g; no decoction. External use: grind into fine powder and apply.

Dianella ensifolia (L.) DC.

Common name: Swordleaf dianella

Property: Perennial herb.

Habitat Distribution: Grown under forests, on hillsides or in grasses; distributed from tropical Asia to Madagascar, Africa.

Medicinal Part: Rhizome.

Efficacies: Drawing out toxin and dispersing swelling.

Indications: Abscesses, ulcerative carbuncles, tinea, lymphadenitis.

Usage and Dosage: This herb, which is extremely poisonous, is strictly forbidden for oral administration. External use: smash with appropriate amount, wrap in gauze and apply on the affected part.

● Araceae

Alocasia cucullata (Lour.) Schott

Common name: Chinese taro

Property: Upright herb.

Habitat Distribution: Grown on creek and valley wetlands or beside fields; cultivated in courtyards or medicinal gardens in some places; distributed in China, Bangladesh, Sri Lanka, Myanmar and Thailand.

Medicinal Part: Whole plant.

Efficacies: Clearing heat and removing toxicity, subduing swelling to relieve pain.

Indications: Influenza, high fever, tuberculosis, acute gastritis, gastric ulcer, chronic gastropathy, ileotyphus; venomous snake bites, cellulitis, sores and furuncles, rheumatism (external use).

Usage and Dosage: This herb, which is toxic, needs to be decocted through decocting for over 6 hours before oral administration to avoid poisoning.

Alocasia macrorrhiza (L.) Schott

Common name: Giant alocasia

Property: Large evergreen herb.

Habitat Distribution: Grown on tropical rainforest edges or under valley musagroves; distributed from Bangladesh, northeast India to the Malay Peninsula, the Indo-China Peninsula, the Philippines and Indonesia.

Medicinal Part: Rhizome.

Efficacies: Clearing heat and removing toxic substances, subsiding a swelling and disintegrating a mass, removing necrotic tissues and promoting regeneration of new tissue.

Indications: Abdominal pain, cholera and hernia,tuberculosis, rheumatoid arthritis, bronchitis, influenza, typhoid, rheumatic heart disease; furuncles and sores, snake and insect bites, burns and scalds (external use); neurodermatitis (external use by mixing with kerosene); bovine cold and swine erysipelas (veterinary use).

Usage and Dosage: Oral administration: make decoction (dried herb 9 to 15 g, fresh herb 30 to 60 g). This herb, which is toxic, must be decocted for a long time to change water twice to thrice before it can be taken. External use: smash appropriate amount of fresh herb and apply on the affected part. It cannot be applied on healthy skin.

● Dioscoreaceae

Dioscorea bulbifera L.

Common name: Air potato yam

Property: Twining herbaceous liana.

Habitat Distribution: Grown on river valley edges, shady valley gullies or mixed forest edges and sometimes around houses or under the tree shades on roadsides; distributed in Japan, North Korea, India, Myanmar, Oceania and Africa.

Medicinal Part: Tuber.

Efficacies: Clearing heat, dispersing swelling and removing toxicity, dissipating phlegm and resolving masses, cooling blood for hemostasis.

Indications: Thyromegaly, lymphatic tuberculosis, swollen sore throat, hematemesis, hemoptysis, pertussis, thyroid tumor, cough and asthma due to excessive phlegm, scrofula, sores, venomous snake bites; sores and boils (external use).

Usage and Dosage: Oral administration: make decoction, 4.5 to 9 g. External use: smash and apply or grind into fine powder and apply.

● Arecaceae

Areca catechu L.

Common name: Betel-nut palm, arecan nut palm, catechu

Property: Arbor.

Habitat Distribution: Widely cultivated in tropical Asia.

Medicinal Part: Fruit.

Efficacies: Destroying parasites, expelling retained food, descending adversely risen Qi to remove stasis, promoting diuresis to resolvedampness.

Indications: Infections of parasites such as tapeworm, hookworm, roundworm, pinworm and fasciolopsis.

Usage and Dosage: Oral administration: make decoction, 90 to 120 g.

Pandanaceae

Pandanus tectorius Sol.

Common name: Pandan, Pandanuss, Thatch screwpine

Property: Evergreen branching shrub or dungarunga.

Habitat Distribution: Grown on seaside sands or introduced as hedgerows; native to the tropical and subtropical areas in Africa, also distributed in tropical Asia and South Australia.

Medicinal Part: Leaf, root and fruit.

Efficacies: Lowering blood glucose, inducing sweating to relieve exterior syndrome, clearing heat and removing toxicity, promoting diuresis, relieving alcoholism.

Indications: Cold fever, nephritis, edema, pain in the loin and lower extremities, liver-heat due to fire of deficiency type, cirrhosis ascites, heat stroke, urinary tract infection, calculus, hepatitis, conjunctivitis, dysentery, orchitis, hemorrhoids.

Usage and Dosage: Oral administration: make decoction, 3–6 g.

● Taccaceae

Tacca leontopetaloides (L.) Kuntze.

Common name: Polynesian Arrowroot

Property: Perennial herb.

Habitat Distribution: Native to Asia and Africa; widely distributed from the western coast of Africa to eastern Pacific.

Medicinal Part: Tuber.

Efficacies: Clearing heat and removing toxicity, regulating Qi-flowing and relieving pain.

Indications: Gastropathy, diarrhea and dysentery, gastrorrhagia and colonorrhagia.

Usage and Dosage: Oral administration: make decoction, 9 to 15 g. External use: smash with appropriate amount and apply. Overdose is forbidden for oral administration. It is contraindicated in pregnancy.

● Orchidaceae

Arundina graminifolia (D. Don) Hochr.

Common name: Grassleaf arundina

Property: Perennial herb.

Habitat Distribution: Grown on grass slopes, near valleys, under shrubs or in forests; distributed in China, Nepal, Sikkim, Bhutan, India, Sri Lanka, Myanmar, Vietnam, Laos, Cambodia, Thailand, Malaysia, Indonesia, the Ryukyu Islands and Tahiti.

Medicinal Part: Whole herb.

Efficacies: Clearing heat and removing toxicity, dispelling wind and removing dampness, relieving pain , promoting diuresis.

Indications: Hepatitis, arthralgia, aching lumbus and leg pain, stomachache, stranguria, dysuria, beriberi edema, scrofula, tuberculosis, toothache, swollen sore throat, cold, infantile convulsions, infantile malnutritional stagnation, cough, food poisoning, traumatic injuries, snake bites, traumatic bleeding.

Usage and Dosage: Oral administration: make decoction, 15 to 30 g; or stew with meat.

● Cyperaceae

Cyperus difformis L.

Common name: Difformed galingale

Property: Annual herb.

Habitat Distribution: Grown in paddy fields or humid places along waterways; distributed in China, Rasia, Japan, North Korea, India, Himalayas, Africa and Central America.

Medicinal Part: Whole herb.

Efficacies: Promoting the circulation of Qi, invigorating the circulation of blood, relieving stranguria, inducing urination.

Indications: Heat strangury, dysuria, traumatic injuries, hematemesis.

Usage and Dosage: Oral administration: make decoction (fresh herb 15 to 30 g); or scorching before grinding into fine powder.

Cyperus rotundus L.

Common name: Jinmen Galingale

Property: Herb.

Habitat Distribution: Grown in the grasses of hillside wastelands or humid places by waterways; widely distributed around the world.

Medicinal Part: Rhizome.

Efficacies: Regulating Qi-flowing to alleviate mental depression, regulating menstruation to relieve dysmenorrhea.

Indications: Liver-Qi stagnation, distending pain in chest and hypochondrium, hernia pain, breast distending pain, Qi stagnation of the spleen and stomach, abdominal congestion, distention and pain, irregular menstruation, amenorrhea and dysmenorrhea.

Usage and Dosage: Oral administration: make decoction, 4.5 to 9 g; or use in pills or powder. External use: grind into fine powder and spread or apply; or make into pies for hot compress.

Kyllinga brevifolia Rottb.

Common name: Samll-star spike kyllinga

Property: Perennial herb.

Habitat Distribution: Grown on hillside wastelands, roadside grasses, grass-spots by the field, streamside and seaside beaches; distributed in the tropical regions of western Africa, Malagasy, Himalayas, China, India, Myanmar, Vietnam, Malaya, Indonesia, the Philippines, Japan, the Ryukyu Islands, Australia and America.

Medicinal Part: Whole herb.

Efficacies: Dispelling wind and relieving exterior syndrome, clearing heat eliminating dampness, relieving cough and reducing sputum, removing stasis and dispersing swelling.

Indications: Cold due to wind-cold, cold-heat headache, arthralgia and myalgia, cough, malaria, jaundice, dysentery, pyogenic infections of ulcers and festers, traumatic injuries and knife wounds.

Usage and Dosage: Oral administration: make decoction (fresh herb 30 to 60 g); or crush to extract juice. External use: smash and apply.

Kyllinga triceps Rottb.

Common name: Threehead kyllinga

Property: Perennial herb.

Habitat Distribution: Grown under hillside forests, on ditch edges, in places close to the water of paddy fields and humid places in open land; distributed in Himalayas, China, India, Myanmar, Thailand, Vietnam, Malaya, Indonesia, the Philippines, Ryukyu Islands, Australia and tropical America.

Medicinal Part: Whole herb.

Efficacies: Ventilating lung-Qi and relieving cough, clearing heat and removing toxicity, removing stasis and dispersing swelling, destroying parasites and preventing malaria.

Indications: Cold cough, pertussis, swollen sore throat, dysentery, venomous snake bites, malaria, traumatic injuries, cutaneous pruritus.

Usage and Dosage: Oral administration: make decoction, 30 to 60 g. External use: smash with appropriate amount and apply; or wash with decoction.

● Poaceae

Cymbopogon citratus (DC.) Stapf

Common name: Lemongrass

Property: Perennial dense scented herb.

Habitat Distribution: Widely planted in tropical regions and also cultivated in the West Indies and East Africa.

Medicinal Part: Leaf.

Efficacies: Dispelling wind to relieve exterior syndrome, ctivating blood to collaterals, promoting diuresis to subduing swelling.

Indications: Cholera, acute gastroenteritis and chronic diarrhea, stomachache, abdominal pain, headache, fever and headache, herpes.

Usage and Dosage: Oral administration: make decoction, 6 to 15 g. External use: wash with decoction of appropriate amount, or grind into fine powder and apply.